advanced

physics

laboratory

book

advanced **physics** *laboratory* book

peter warren

JOHN MURRAY

Cover photo: Martin Bond/Science Photo Library

First published by
John Murray Publishers, a member of the Hodder Headline Group
338 Euston Road
London NW1 3BH

Reprinted 2004

Layouts by Stephen Rowling/springworks
Artwork by Barking Dog Art
CD-ROM produced by Peter Warren
Cover design by John Townson/Creation
Cover photo courtesy of Martin Bond/Science Photo Library

Typeset in 11/13 Bodoni Book by Tech-Set Ltd.
Printed and bound in Great Britain by J.W. Arrowsmiths Limited, Bristol

A catalogue entry for this title is available from the British Library

ISBN 0 7195 8054 4

Contents

Introduction

This is a practical course book that I hope will encourage you to experiment. It contains details of experiments that relate to the content of the A level Physics syllabuses.

Physicists use experiments and theory to investigate the natural world. Practical investigation is a powerful tool for making new discoveries and useful inventions. At the end of your Physics course, you should have the skills and confidence to plan and carry out your own experiments. There are good reasons for doing practical work. It develops skills that are generally useful in solving problems. It requires us to process measurements and test hypotheses. This deepens our understanding and helps us to learn. It teaches us to be objective and come to conclusions based on evidence.

This is also a coursework book. It contains details of how to plan, analyse and evaluate an experiment. It will help you to produce an investigation that meets the highest mark descriptors of the examining boards.

Experiments are seen as having four stages: *planning*, *implementing*, *analysing*, and *evaluating*. **Section A** of this book contains suggestions for each of these stages, and can be used to develop your own investigations.

Section B describes a wide range of experiments and investigations, on nearly every topic in the A level course. For each experiment, the *Plan* gives sufficient information for you to obtain a set of readings without coming to any harm. The *Analysis* suggests how the readings can be processed to reach a clear conclusion. The little pens indicate written tasks for you to do. The *Evaluation* helps with the difficult task of estimating the uncertainties in the readings.

Skill level boxes allow you to grade your implementing and analysing skills.

Sample readings are included as examples of how to lay out results. They can also be used as exercises in processing data, and for this purpose some sets of readings, indicated by , are included as spreadsheet files on the accompanying CD-ROM (CD).

The **computer simulations** on the CD are experiments that you can do with a mouse. You can control variables, take readings and plot graphs. The simulation programs mostly illustrate experiments that are difficult to do in the school lab. There are Help files with notes on the experiments, suggestions for changing variables, as well as definitions and theory. Brief instructions and suggestions for analysis are given in this book. These are also indicated by . The programs are also available at:

www.physicslab.co.uk

Working through these programs will complement the practical work and will help you to see how simple things are when you understand them!

Peter Warren

Running the CD

Minimum System Requirements

- Pentium PC
- Windows 95 or above
- 2 Mb RAM (9.5 Mb available hard-disk space if installing the programs)
- 1024 x 768 screen resolution
- Microsoft Excel 5.0/95 Workbook or above
- Internet access (to download program updates)

If the CD does not run automatically, locate the 'PhysicsLab' folder, and click on 'PhyLabindex.exe'.

Acknowledgements

I would like to thank Veronica Quainoo, Senior Technician at Grey Coat Hospital School, Westminster, and Tom Chipchase, Daisy French, John King, Catharine Lau, Emma O'Hanlon, Molly Song and Portia Woche, students at the school, who helped me to test the experiments and obtain the sample results. I am also very grateful to Jane Roth for her detailed and expert editorial work on the manuscript and for her many suggestions for improvements to the programs.

Peter Warren

SECTION

A Let's investigate

Why experiment?

We experiment to find out, to measure things and to test theories. It is more interesting to find out than to be told – and perhaps no one knows the answer, anyway. We have to think for ourselves when we set up an experiment. We have to work with apparatus, control variables and gather data. We have to be inventive and open-minded, and we have to co-operate with others.

The scientific way is to gather evidence that convinces ourselves, and then others, of our theories. The skills we develop as we experiment help us to become good problem-solvers.

This section of the book deals with the four stages of experimental work: *planning*, *implementing*, *analysing*, and *evaluating*.

1 Planning

Taking precise and accurate measurements

Experimenting in physics nearly always means taking measurements. We use the numbers to look for patterns or to calculate physical values. The main measurements are length, time, mass, potential difference (pd), current and temperature, and sometimes we have to count.

Experiments should be designed to give measurements that are precise and accurate.

What we mean by precise and accurate measurements

Taking a measurement in an experiment is rather like shooting at a target with a rifle. We aim to get a result that is 'spot on'. A skilled marksman, using a rifle with a telescopic sight, will make **precise** shots. The bullet holes will be scattered over a small area around the target centre. See Figure 1.1a. A less skilful person will have shots that are more scattered (less precise). See Figure 1.1b.

An **accurate** measurement is one that is close to the 'true' value. If the alignment of the marksman's telescopic sight on a rifle is faulty, his shots will still be precise but will be off target and not accurate. See Figure 1.1c.

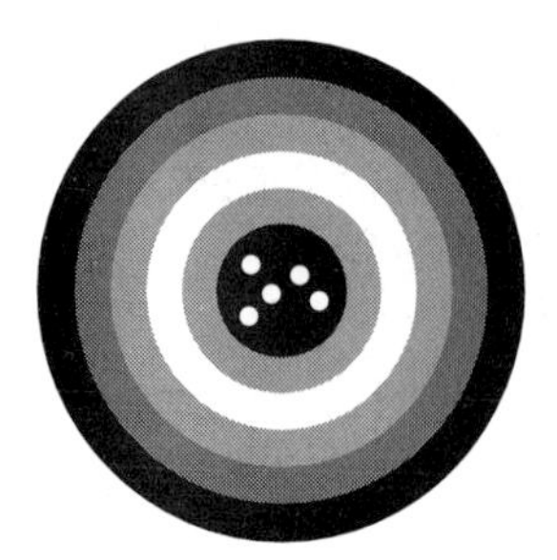

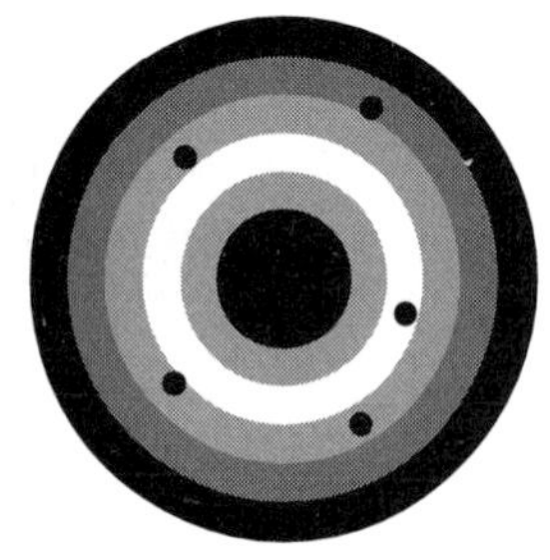

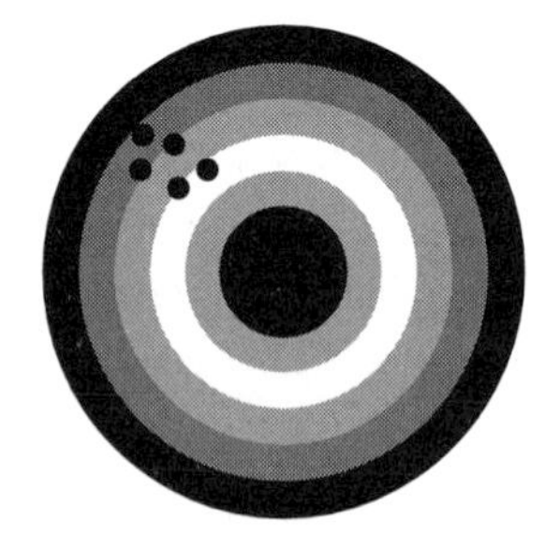

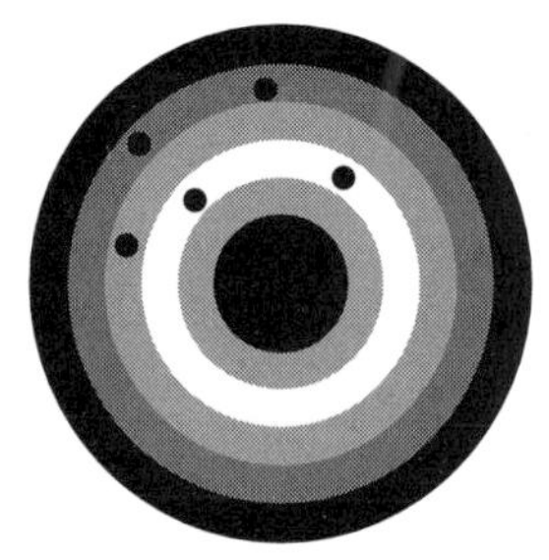

Figure 1.1

a Precise and accurate
Example: measuring mass with an electronic balance. The random uncertainties are small. The systematic error is small.

b Imprecise but accurate
Example: measuring the speed of sound by echoes (p. 142), or measuring wavelength by two-slit interference (p. 157). The random uncertainties are large. The systematic error is small.

c Precise but inaccurate
Example: measuring the diameter of a ball bearing with a micrometer that has a zero error, or using a ticker timer to measure speed (p. 24). The random uncertainties are small. The systematic error is large.

d Imprecise and inaccurate
Example: measuring mass using bathroom scales with a zero error. The random uncertainties are large. The systematic error is large.

To get *precise* readings, an operator needs to use an exact measuring instrument with skill and with a sound technique.

To get *accurate* readings, the instrument must be correctly adjusted and calibrated. Also, the instrument or the design of the experiment must not alter the quantity that is being measured.

For example, for the measurement of the diameter of a small steel ball to be *precise*:

- use a micrometer (an exact instrument);
- use the ratchet to tighten the jaws onto the ball (a sound technique);
- read the scale correctly (skill).

To make sure the reading is *accurate*:

- close the jaws and check for a zero error (correct adjustment).

Improving precision

To make a more precise measurement of the extension of a copper wire under a tensile force (p. 61), for example, we could do the following.

Increase the exactness of the instrument.	Use a vernier scale that reads to 0.1 mm rather than a metre rule that reads to 1 mm.
Increase skill.	Take a lesson from the teacher in reading a vernier and then practise.
Think out an effective technique.	Use several metres of wire to increase the extension and use a lens to read the vernier.

The cost of improving precision is that it will probably take longer and be more expensive to take a measurement. Precision does however give a comforting reliability to the data by reducing random uncertainties, although there is no point in making one measurement precise if another related quantity is measured with poor precision.

Improving accuracy

Zero errors in instruments, wrongly calibrated scales and faulty techniques will make readings inaccurate. These are sources of 'systematic error' that can be corrected until all distracting effects have been removed and the measurement is as accurate as possible.

A zero error should be reset if there is an adjusting screw, or deducted from the measurements.

Scales should be checked by comparing readings made with identical instruments. Two ammeters can be connected in series or two voltmeters in parallel to check for differences. The scales of thermometers, metre rules and stop watches can be compared by using them in pairs to measure the same quantity.

Some faulty techniques to be wary of are:

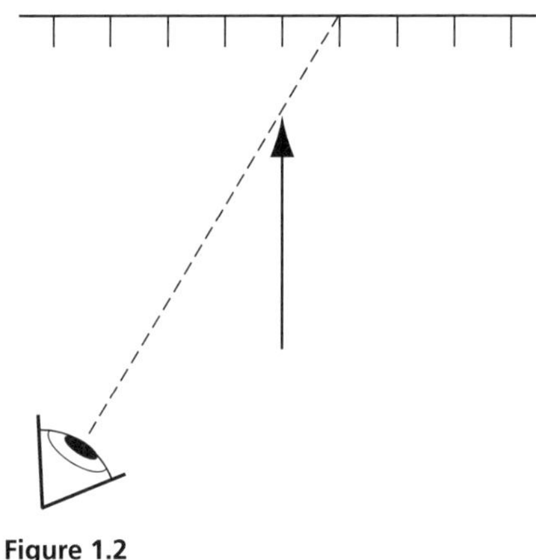
Figure 1.2

- parallax when reading a scale – the pointer, scale and eye should be in line, to avoid a parallax effect as shown in Figure 1.2;
- miscounting when timing oscillations;
- 'back-lash' when a screw thread is used to adjust an instrument – it should be moved to its measuring point by turning the screw against a spring, rather than relying on a spring to move it back;
- over-tightening a micrometer;
- not stirring when reading the temperature of a liquid;
- slow reaction times when using a stop watch.

Blunders (or anomalies)

A measurement may be wrong because a mistake was made when it was taken. If the measurement technique is precise, the blunder will be more obvious and can be left out of further calculations or ignored when drawing a line on a graph. Blunders are often first spotted on graphs.

Planning an investigation

Step 1 Define the problem and make a clear aim

The topic for investigation is often given as a general instruction. The first step is to use what you know and understand about science to define the exact problem you are going to investigate. You can then go on to form your own prediction and a clear aim for the test. See Figure 1.3.

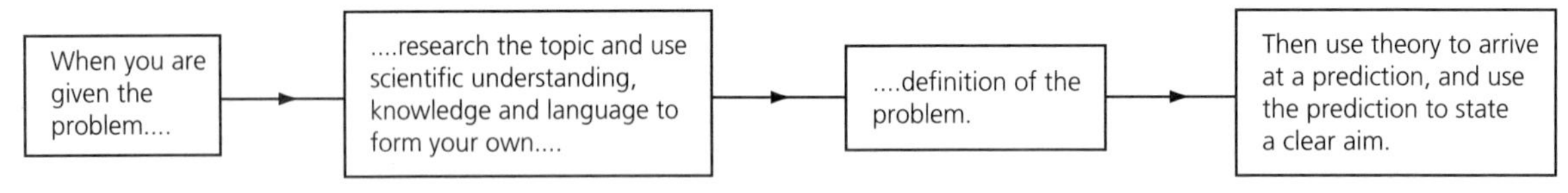

Figure 1.3

Here are suggestions for investigations, showing this first step.

The given problem	My defined problem	My prediction and aim
Investigate the discharge of a capacitor through a resistor (p. 96).	To find how the pd across the capacitor changes as it discharges through a resistor.	The discharge is exponential. My aim is to show that a graph of ln (pd) against t is a straight line.
Investigate the resistance of a thermistor (p. 80).	To find out if a thermistor follows Ohm's law at different temperatures.	It will follow Ohm's law. I aim to show that graphs of pd against current are straight lines through the origin for a fixed temperature.
Investigate the force on a current in a magnetic field.	To find out how the force on a coil of wire depends on the length of the wire in the field and on the current.	The force will be proportional to the length and the current. My aim is to check this proportionality.
Investigate the stretching of springs (p. 59).	To find the elastic limit and the spring constant of a spring.	The spring will follow Hooke's law up to a certain limited extension. My aim is to measure the elastic limit and the spring constant.
Investigate the vertical oscillations of a mass on a spring (p. 130).	To find out how the period of oscillation of a spring depends on the mass it is carrying.	The motion will be SHM. A graph of $(\text{period})^2$ against mass should be a straight line. My aim is to test this relationship.
Investigate the energy needed to heat materials (pp. 169, 171).	To compare the specific heat capacities of a metal (copper), a liquid (cooking oil), and water.	It will require more energy to heat the water than the oil and the metal. My aim is to find a numerical comparison.
Investigate the passage of gamma rays through air (p. 194).	To find out whether gamma rays from a point source follow the inverse square law.	The gamma ray count will fall off as the square of the distance from the source. My aim is to show that a graph of count rate against $1/(\text{distance})^2$ is a straight line.
Estimate the absolute zero of temperature (p. 179).	To measure the expansion of dry air with increasing temperature, and find the temperature at which its volume would be zero.	The volume should increase uniformly with temperature. My aim is to produce a graph of volume against temperature and extrapolate it back to zero volume to get the absolute zero of temperature.
Investigate the internal resistance of batteries (p. 85).	To measure the internal resistance of 1.5 V cells of different sizes.	The internal resistance should decrease with the cross-sectional area and increase with the length of the cells. My aim is to compare the resistivity of the cells.

Step 2 Choose the apparatus

From your research work you will have an experiment in mind. The next step is to choose the best instruments for taking the measurements. From those available, you will have to make choices depending on the precision needed. See the table below for some examples.

Quantity to measure	Instruments available	Smallest division or range
angle	protractor	1°
length	metre rule vernier callipers micrometer	1 mm 0.1 mm 0.01 mm
volume	100 cm^3 measuring cylinder 10 cm^3 measuring cylinder	1.0 cm^3 0.2 cm^3
current	analogue ammeter digital ammeter	1 division (depending on the range) 0.02 A
voltage, current, resistance	multimeter	depends on range selected
varying voltage	oscilloscope or datalogger	μV or mV depending on the instrument
temperature	standard liquid-in-glass thermometer 'special range' liquid-in-glass thermometer electronic temperature probe with computer interface	1 °C 0.5 °C (depending on the range) usually the same precision as a digital meter
time	digital stop watch or clock oscilloscope time base or datalogger	0.1 s can be as small as 1 μs
emf	continuously variable, smoothed and regulated power supply stepped power supply	steady emf that can be adjusted in very small steps stepped emf (usually dependent on the current drawn)
light intensity	LDR with datalogger or suitable circuitry	sensitivity depends on the instrument
mass	electronic balance	0.1 g or 0.01 g

Step 3 Think about safety

Tall structures, including beakers of hot liquids on tripods, can fall over. Stressed wires can snap and heavy weights can fall on feet. Bunsen flames can set light to long hair and flammable liquids. Hot tripods and beakers can cause nasty burns. High voltages can shock and burn the body. Large currents through wires can cause burns and ruin sensitive meters. Radiation from radioactive sources and UV lamps can damage cells in the body.

Your plan should point out such hazards and include instructions that reduce the risk of harm to people and damage to equipment.

Step 4 Initial plan of action

Outline the plan that you think will achieve the results you need to test your prediction. The plan should be in a series of 'steps' and include all the small details that ensure the readings will be reliable and can be measured safely. It is useful to include an apparatus list, with an explanation against each instrument saying why it has been chosen for the job.

Step 5 Test run

Before you start taking careful readings, it is important to check that the apparatus works and that the instruments give readings of a suitable size. Assemble the apparatus and alter the independent variable to find the maximum and minimum values of the readings. From these decide how many readings you are going to take and the intervals between them. For example, if a change of pd from 0 to 10 V gives a rise in current from 0 to 2 A you could choose to take a reading every 1 volt, giving 10 readings altogether. Remember it is better science to take readings closer together over the part of the range where quantities change more rapidly (see p. 13).

Step 6 Final plan of action

Having used the apparatus, you will be able to add practical hints to the plan that ensure your experiment will produce precise and accurate data.

2 Implementing

Techniques

To do an experiment well requires skill. You need good manual skills; you have to be good at controlling variables and good at reading instruments. Skill comes with practice, but here are some commonsense techniques that help with obtaining precise measurements.

1 Timing oscillations (e.g. loaded spring, pendulum bob, torsional oscillations of a metre rule)

- Time from the centre of the oscillation where the movement is fastest.
- Have a (fiducial) mark behind the object that can be used to judge when the oscillation passes the centre. Count each time it passes the mark in the *same* direction.
- Start the oscillation before you start timing and let spurious minor oscillations die out.
- Count 0 as the object passes the mark and start the stop watch. When it next passes the mark count 1, and so on.
- Time enough oscillations to get a total time of at least 30 s. This will mean timing more oscillations when the period is short.

2 Heating a liquid in a beaker with a Bunsen burner and measuring its temperature

- Use tripod, gauze, bench mat and safety glasses. Tie your hair back if it is long. Clamp equipment that could fall over.
- If possible, use an electronic thermometer with a digital read-out.
- Use a blue flame of medium size.
- Put the Bunsen under the beaker until the temperature has risen by a suitable amount, say 10 degrees.
- Remove the Bunsen and stir the liquid with a stirrer until the temperature stabilises.
- Leave the thermometer in the liquid. If using a mercury thermometer, get down level to read the top of the mercury meniscus.

- Make sure the leads of an electronic thermometer cannot be damaged by the Bunsen flame or hot gauze.
- Allow equipment to cool afterwards before handling it. Always lift a tripod by the bottom of its legs.

3 Taking moving readings (e.g. the pd of a discharging capacitor or counts from a decaying radionuclide)

- Use a digital voltmeter or counter and fix your eyes on the display while a partner indicates when to take the reading.
- Use the lap-timing facility of a stop watch. Read the instrument and press the lap timer at the same moment. This freezes the time for a few seconds while you record the watch and instrument readings. The watch display then goes back automatically to the elapsed time.
- For rapidly changing quantities use a datalogger and computer (see p. 12).

4 Measuring direct pd and current

- Use a smoothed, regulated power supply (or a battery).
- With analogue meters, vary one quantity so that the pointer lies exactly on a scale division. Then only one meter is awkward to read.
- With digital meters, choose the range that gives the largest number of decimal places. Don't change the range in the middle of an experiment because this may alter the impedance of the instrument.
- Do not always increase pd in equal steps. Take more readings when the current changes rapidly.
- The pd of the power supply may drop as it delivers more current. Don't assume it will be constant.
- If it is important to keep the temperature of a component constant, switch on briefly while you take the reading, then switch off to allow the component to cool.

5 Using analogue meters

- Check the pointer reads zero when there is no current. If it doesn't, adjust the screw at the base of the pointer until it does.
- Position your eye vertically above the needle. If there is a strip of mirror behind the pointer, move your eye until the needle and its image coincide.

6 Building electric circuits

- Connect the components together with the power switched off and without the voltmeter.
- Connect the voltmeter across the component being tested.
- Use 4 mm plugs and screw-down terminal blocks to make connections electrically secure. Only use crocodile clips as a last resort.
- Check polarities. For ammeters and voltmeters, it's 'red-to-red-or-you're-dead'!
- With multimeters, put them on their highest voltage or current range and adjust down to more sensitive ranges in stages.

- If you are not a confident circuit builder, have the circuit checked before you switch on.
- Turn the power supply down to zero before you switch on. (Remember, even in low voltage experiments, there are high voltages about. Power packs use the mains supply, and switching off currents in large coils can induce thousands of volts.)

7 Measuring stretch (e.g. springs, elastic bands, copper wire)

- Make the marker or the end of the spring lie on the metre rule.
- Do not try to make the zero of the scale coincide with the marker. Take the reading with zero pulling force as the initial reading.
- Record actual readings as force is applied. Calculate the extension later in a results table. Do not do sums in your head.
- Take readings 'loading' and 'unloading', in case there is a difference in behaviour.
- Protect your eyes with safety glasses and the rest of your body by staying away from the danger zone. (The energy stored in stressed materials can cause fragments to fly at high speed when they snap.)

8 Measuring the diameter of round objects

- Diameters > 5 cm, e.g. tennis ball. Use a metre rule and two set squares. Record the positions of both verticals and subtract them later. Alternatively, open callipers to the size of the object and read off the distance between the points on a metre rule.

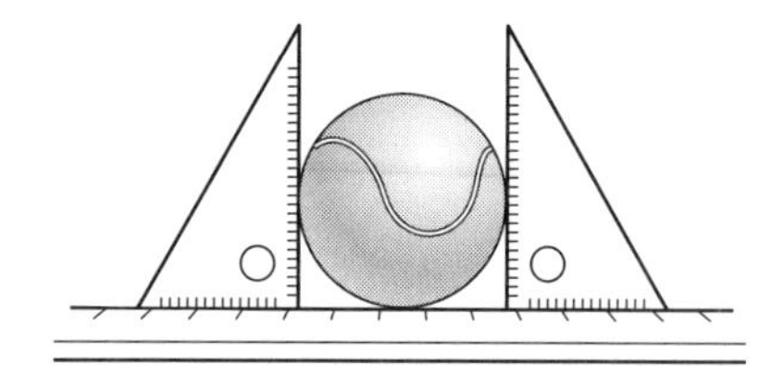

Figure 2.1

- Diameters between 5 mm and 5 cm, e.g. test tube. Use vernier callipers to grip the object, and read the diameter from the scale and vernier. It should be read to 0.1 mm. One check reading should be enough if it gives the same result.

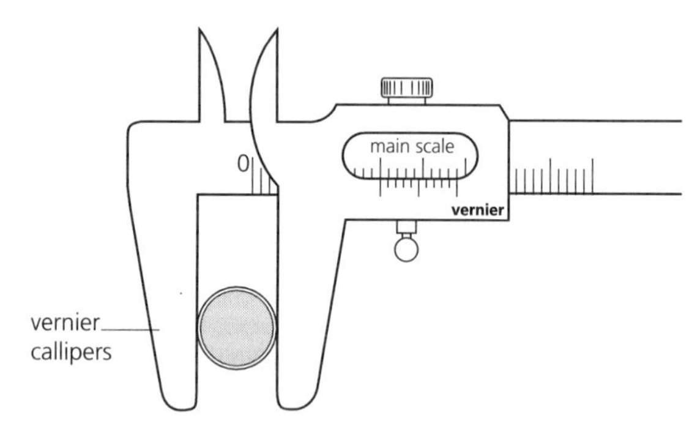

Figure 2.2

- Diameters less than 5 mm, e.g. small ball bearing or diameter of bare wire. Use a micrometer screw gauge. First close the jaws and check for a zero error. Record any zero error and decide whether it should be added or subtracted from readings. Then pinch the object in the jaws using the ratchet and take the reading.

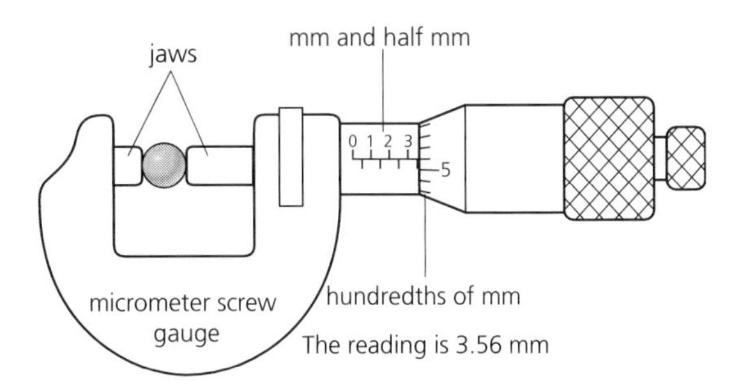

Figure 2.3

9 Measuring lengths

- Use a ruler to measure lengths such as the distance between dots on a ticker tape or the dimensions of a rectangular block. A ruler has a bevelled edge and so the scale markings are close to the object. Transparent rulers should be used to draw best-fit straight lines on graphs.
- To use a metre rule as a scale, position it so that it is as close to the object as possible. Position yourself so that you are in a comfortable viewing position with your eye perpendicular to the scale. Record positions on the scale and do subtractions later in a table. A metre rule should be read to 1 mm.
- Small extensions of elastic materials such as copper or nylon should be measured with a scale and vernier that can measure to 0.1 mm.

10 Using an oscilloscope

- Switch on the instrument and find the spot or line that the electron beam draws on the screen. Helpful tips are to turn up the brightness; press all buttons out (except 'on'); turn the Y-amplifier setting to its highest volts/cm value; turn the 'trigger' off; vary the X- and Y-shift controls.
- Focus and centre the trace. Turn the brightness down until you have a fine line.
- Connect your signal voltage to the input socket.
- Adjust the time base and Y-amplifier sensitivity until you have a stable picture of the signal voltage.

11 Using a computer to take measurements

You will need:

1 An analogue to digital converter (ADC) that plugs into the parallel port or serial port or USB port of the computer. This unit converts analogue voltages from a sensor into a digital signal that can be read by the computer. It is often called an 'interface unit'.
2 A sensor that can be plugged into the ADC.
3 The ADC's program running on the computer to capture and process the values. Detailed instructions of how to work the program will be given in its Help file or instruction leaflet.

The sensor, the ADC, its software and the computer can together be called a 'datalogger'. A range of dataloggers is available but it is worth noting that they quickly become superseded by improved versions. A datalogger can be used for the following.

- *Measuring and storing voltage readings.* This is useful if the voltage in a circuit is changing with time, as with the discharge of a capacitor through a resistor (p. 96) or the voltage waveform from a microphone. The interface unit will have sockets or screw terminals that connect to your circuit in place of the voltmeter.
- *Measuring temperature.* A temperature probe is useful for taking readings automatically, for example when measuring specific heat capacity or heating water in a kettle (pp. 169, 176). The electronic probe may have its own meter that plugs into the interface unit.
- *Measuring speeds.* A light sensor can be used to record the shadow of a passing trolley (p. 28). It is best to shine a small torch at the sensor so that there is a sharp change in the reading when the trolley blocks the light. This arrangement is sometimes called a 'light gate'. The time of passage of the trolley can then be deduced from the readings.
- *Measuring light intensity.* A small light sensor can be moved steadily across interference or diffraction patterns to plot an intensity graph of the pattern (p. 156).
- *Measuring magnetic flux density.* A Hall probe can be moved steadily through a magnetic field and a graph of magnetic flux density shown on the computer screen.
- *Displaying voltage/current characteristics.* You will need to plug voltage- and current-measuring units into the interface. Connect them into your circuit with a variable power supply. Start recording and then increase the pd steadily and gradually over the range you have chosen. The pd and the current will be measured against time. If the software has a 'snap-shot' facility, you can record individual readings of pd and current as you progress through the range. You can then plot the voltage/current graph from the data after you have finished.
- *Measuring resistance.* Resistance terminals and a temperature probe, plugged into the interface, will allow you to automatically measure the resistance and temperature of a thermistor as it cools in hot water.

The range and interval of readings

The apparatus will determine the range of possible readings. The pd applied to a 12 V 24 W lamp, for example, should only be varied between 0 and 12 V. The distance you can move a GM tube from a gamma source is limited to where the count drops to the background level. The mass you can bounce on the end of a spring is limited by the elastic limit of the spring. The length of copper wire you can stretch is limited by the size of the laboratory.

You should use as large a range as is sensible and you should make sure that your readings cover the whole range. Usually readings can be taken at regular intervals across the range. Sometimes, however, readings should be taken closer together over part of the range, when change is happening rapidly. To get a good pd/current characteristic for a lamp filament, for example, you should take current readings at smaller pd intervals when the voltage is low because the graph is more curved over that region. This will give more points for drawing an accurate line.

Repeat readings

A sensible rule is to make one 'check' reading. If it is the same or close to the first one they can be averaged. If it is not, a third reading should confirm which value is reliable.

Some quantities should not be repeated because it is not possible to reproduce the same conditions. These include temperature and pd (often set by a wire-wound potentiometer). You cannot be sure you are measuring the same thing twice. Instead you could take readings at closer intervals.

3 Analysing

The analysis of an experiment starts with recording the results, then processing them in some way and finishing off with a conclusion.

Recording readings

Here are some commonsense rules for recording readings.

Using a table

Often the clearest way to record measurements for a report is in a table. This can be done roughly in a lab notebook during the experiment and typed up later for the report.

- The table should have enough columns for the readings and all the calculations.
- It should have column headings describing the quantity and giving its unit. The recommended notation is **mass /kg**. This makes the quantity in the table dimensionless. '1.5 kg' divided by 'kg' gives the number 1.5. Subsequent processing is then dealing with numbers. If you use brackets to enclose the unit, e.g. **pd (V)**, it can be confused with the symbol for the quantity.
- The columns should be in an order that works to the right as processing is done.
- Where possible the table should be fitted onto one page – it is difficult to read and see patterns in split tables.

Many of the experiments described in Section B have sample readings, which give examples of using tables to record measurements.

Recording raw readings

It is safest to write down a reading exactly as it is given by the instrument. This avoids having to do calculations in your head between reading and writing. The number of decimal places of your measurement will then be set by that given by the instrument.

Stop watch readings should be written as 'time /s = 37.28' or 'time /minutes and seconds = 1.08.32' if the time is over 1 minute. Use a second table column for working out the time in seconds. Note that the second decimal place does not represent greater precision, and subsequent calculations should treat the time value as recorded to 1 decimal place, i.e. one-tenth of a second.

A ruler can be read to 1 mm and lengths should be recorded in centimetres, e.g. 'length /cm = 9.4, 19.4, 109.4' to 1 decimal place.

A metre rule used as a scale, to measure for example the extension of a spring, should also be read to 1 mm. The beginning and end of the extension should be recorded and the difference calculated later in the results table.

Digital meters often give values to 2 decimal places and readings should be recorded as they are seen, to the same number of decimal places.

Liquid-in-glass thermometers can be read to 0.5 °C and so temperatures should be recorded in half degrees, e.g. 'temperature /°C = 19.0' or '19.5'. The decimal point is significant because writing '19' could mean the value lies anywhere between 18 and 20, whereas '19.0' means the value lies between 18.5 and 19.5.

Using a spreadsheet

An excellent technique is to enter readings into a computer spreadsheet directly as they are measured.

Set up the column headings before you start and program in the formulae that are needed to do calculations. It may also be possible for the program to plot a graph of quantities as measurements are entered. Anomalous readings can then be spotted immediately and re-measured. The graph will also show where quantities change rapidly and so where to take readings at closer intervals. It is safest to use the 'scatter graph' option without a line and then draw the best-fit line by hand on the printout. This avoids dot-to-dot lines and allows you to exclude anomalous points.

In the example shown in Figure 3.1, from the 'Measuring inertial mass' experiment (p. 45), the number of dots printed by a ticker timer are converted into a time by dividing by 50. The formula in C5 needed to do this is **=B5/50**. The acceleration in E5 is calculated by the formula **=2*D5/(100*C5*C5)** and mass in F5 by **=A5/E5**. These formulae are then 'filled down' to the bottom row (in Excel, by going to Edit–Fill–Down). Lastly, the A and E columns should be selected to plot a scatter graph. The calculations are done as soon as the readings are entered and a point is plotted on the graph. Full instructions for entering formulae can be found in the Excel Help file. (Other spreadsheet programs should have their own 'help' files.)

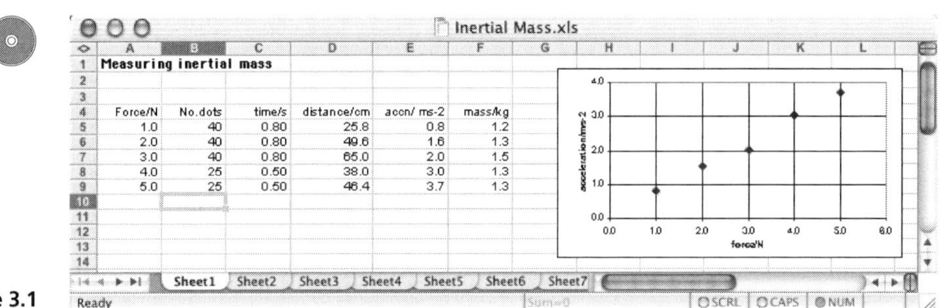

Inertial Mass.xls

Measuring inertial mass

Force/N	No.dots	time/s	distance/cm	accn/ ms-2	mass/kg
1.0	40	0.80	25.8	0.8	1.2
2.0	40	0.80	49.6	1.6	1.3
3.0	40	0.80	65.0	2.0	1.5
4.0	25	0.50	38.0	3.0	1.3
5.0	25	0.50	46.4	3.7	1.3

Figure 3.1

Processing results

When you put your readings into a formula and use a calculator or spreadsheet to get an answer, this will be given to 10 or more figures. Most of these numbers are not significant and the answer must be rounded down to a realistic value. In the example on the previous page, all the measurements were made to 2 significant figures except distance, which was measured to 3. So the calculation of mass is given to 2 significant figures. A sensible rule is to *quote the same number of significant figures in the answer as the smallest number of significant figures in the raw data.*

Averaging

We want our measured value to be the 'spot-on' true value. A set of measurements, however, will be scattered randomly about the unknown true value. If we take an average, readings that are a little too high combine with readings that are a little too low to reduce the variations. The average value is the closest we can get to the true value.

The formula for calculating the average of the masses in the spreadsheet example on p. 15 is **=Average(F5:F9)**.

Standard deviation

Calculation of the *standard deviation* is a convenient way of estimating the uncertainty in a set of measurements or calculations that have been averaged. Calculators and spreadsheets can do the sums for you. In the spreadsheet example on p. 15 the formula is **=STDEVP(F5:F9)**.

Standard deviation is the square root of the average of the sum of the squares of the differences of each reading from the average reading. If the variation in the readings results from random errors, you can be sure that two-thirds of the readings will lie closer to the average than the standard deviation. It is a measure of the spread of your readings about the average and the precision with which they were taken.

The abbreviation used for standard deviation in the sample readings in Section B is 'std'.

Plotting graphs

Graphs give a visual indication of how quantities vary with one another; they show up anomalous values, and gradients can often be used to get an average value using all of the readings. Here are some suggestions for producing precision graphs.

- The scales of the axes should be chosen so that it is easy to plot and read the co-ordinates of the points.
- The plotted values should fill the graph sheet.
- The axes should be labelled with the name and unit of the quantity.
- Points should be plotted accurately with a sharp pencil.

- Best-fit lines should be chosen and drawn as straight lines or smooth curves. It is legitimate to leave out a point if it doesn't fit with all the others. Lines should not be forced through the origin. Dot-to-dot lines are not acceptable because, as a rule, physical quantities do not change abruptly.
- Gradients should be measured by drawing a *large* right-angled triangle and writing the co-ordinates of both ends of the hypotenuse on the graph paper.

Note that in some mathematical teaching the word *line* is used to mean *straight line*. Here the terms *curved line* and *straight line* will be used.

How *not* to plot a graph

The following diagrams illustrate pitfalls to avoid when plotting graphs.

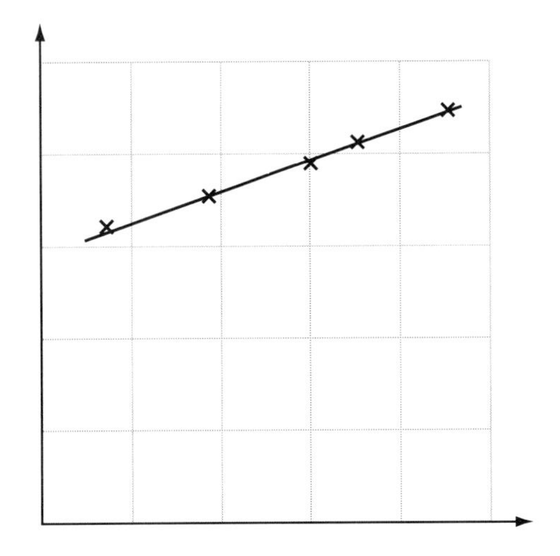

Figure 3.2a Poor *y*-scale. Points fill less than half of the graph paper. Detail from the *y*-readings is lost.

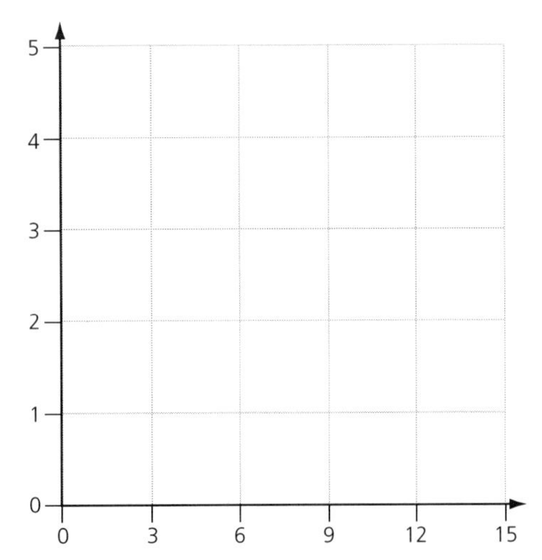

Figure 3.2b Awkward scale divisions on the *x*-axis. They will be difficult to read accurately.

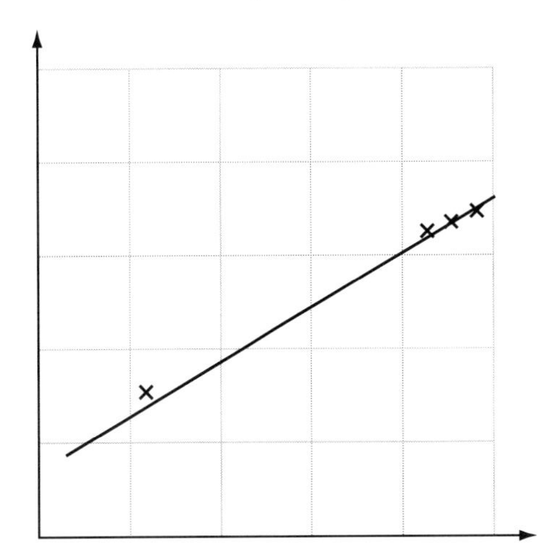

Figure 3.2c Poor distribution of readings makes it difficult to choose a best-fit line.

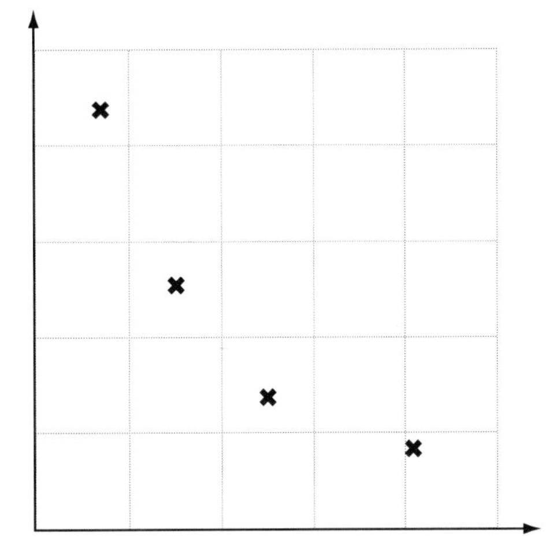

Figure 3.2d The plotted points are too thick, covering too large an area. Precision is lost.

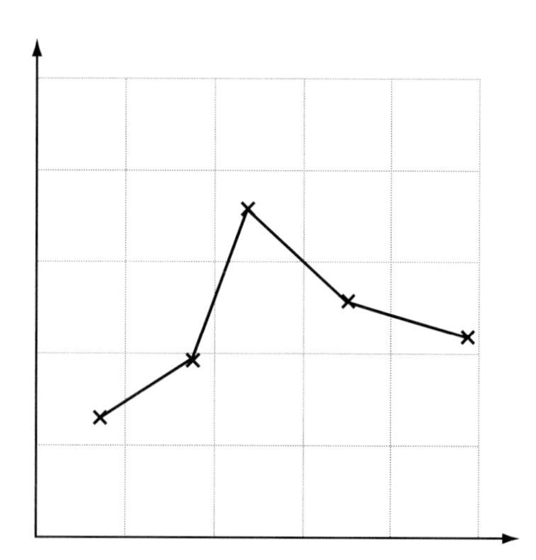

Figure 3.2e The line has been drawn dot-to-dot. Physical quantities do not usually change in this way. The scatter is due to uncertainties in the readings.

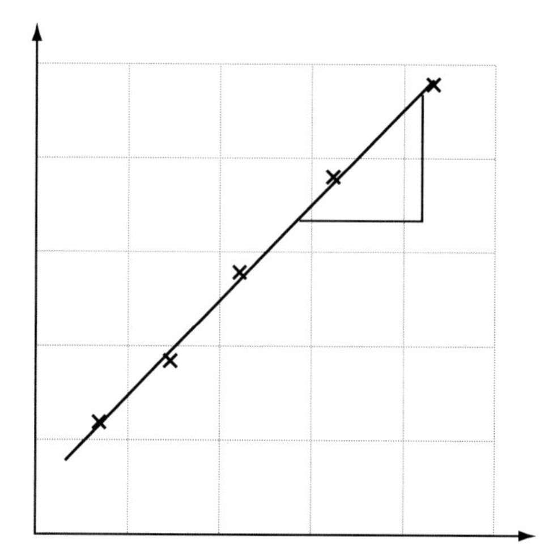

Figure 3.2f The triangle used to calculate the gradient is too small. Large % errors would be introduced when reading the co-ordinates of each end of the hypotenuse.

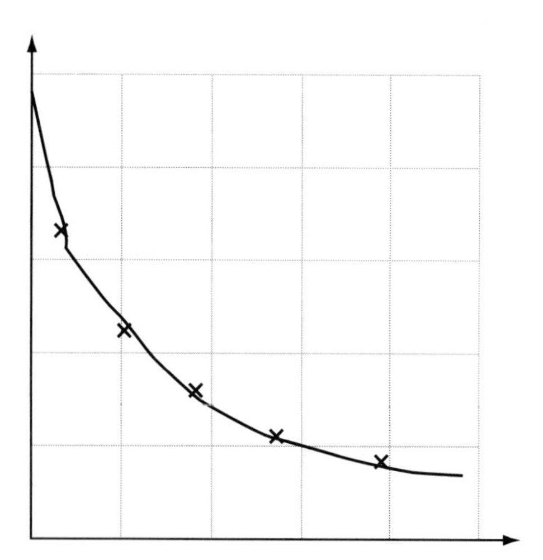

Figure 3.2g The curved line through the points has been drawn badly. It is not smooth and is not the best-fit line.

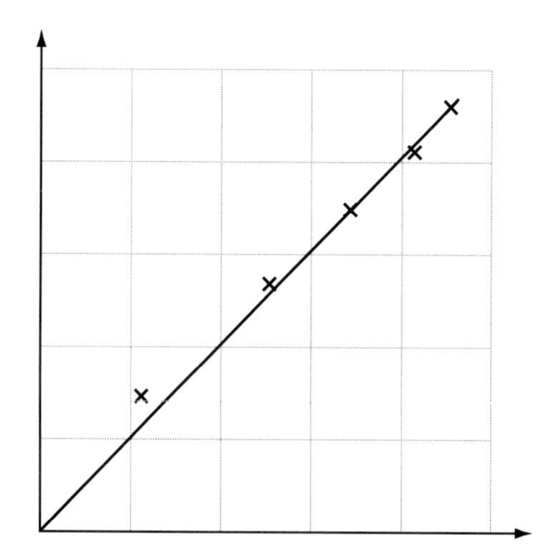

Figure 3.2h The line drawn has been forced through the origin and is not the best-fit line.

The conclusion

The analysis of your readings should allow you to come to a conclusion from your work. It may only be necessary to state that your findings support the original prediction. If there is only partial agreement, or the findings are inconclusive, you will have to put conditions into your summing up. The conclusion should be the final statement of a clear logical process that started with the plan and continued with the collection and processing of the data. It should be brief, expressed clearly and should state the limits within which you think the conclusion is valid (see p. 22). For example, 'The investigation shows that, within the uncertainties of my measurements, the tensile force and the extension of the spring are proportional over the range of the test'.

4 Evaluating

To convince yourself and others of the value of an experiment and of the validity of its conclusion, it is important to identify the limitations that are set by the instruments and the method. An evaluation should estimate the size of the uncertainty in the final result and the limits of accuracy of the concluding statement. This evaluation should lead to suggestions for improvements that would give more precise and accurate readings.

Evaluation of limitations and uncertainties

A full evaluation should contain answers to the following questions:

... about the equipment and method
What were the limitations of the instruments used? How could the instruments be improved?
What were the limitations of the method used to get the readings? How could the method be improved?
Why would the improvements make the results more accurate or more precise or both?

... about the reliability of the measurements
Which measurement introduces the greatest uncertainty?
What is the % uncertainty in each of the measurements?
What is the overall % uncertainty in the final result?
To what degree of accuracy is the conclusion valid, bearing in mind the uncertainties? Or is it valid at all?

Estimating uncertainty in a set of readings that can be averaged

Each reading will have an *uncertainty*, that is, the amount it differs from the 'true value'. Since the true value is not known, the uncertainty cannot be known exactly. The best we can do is to estimate its size.

Method 1 Find the average of the set of readings.
Work out the numerical differences from the average.
Find the average of these differences.

Method 2 Find the standard deviation using a spreadsheet or a calculator.
Enter the code and the numbers, then let the program do the calculation (see p. 16).

Either method will give an acceptable value for the uncertainty.

Sample readings The masses of 1 N weight hangers were measured by an electronic balance.

mass of 1 N weight hangers /g		101.92	101.76	101.36	100.96	100.57	101.18	100.42
numerical difference from average /g		0.75	0.59	0.19	0.21	0.60	0.01	0.75
average mass /g	**101.17**							
average of differences /g	**0.44** (method 1)							
standard deviation /g	**0.52** (method 2)							
estimate of uncertainty /%	**0.5** (rounded down because it is only an estimate)							

The small calibration error of the electronic balance, supplied by the manufacturer, would have to be added to give the overall uncertainty (p. 22). This is an example of a precise and accurate experiment!

Note that the formula to enter in an Excel spreadsheet to calculate the standard deviation of a set of numbers is:

=STDEVP(number1,number2,....)

Estimating uncertainty in a single reading

A sensible rule of thumb is that *the uncertainty in a single reading is half the smallest division of the instrument used.*

This would give uncertainties as follows, for example.

standard laboratory thermometer	0.5 °C
metre rule	0.5 mm
micrometer	0.005 mm
analogue meter with 0.1 divisions	0.05

The rule would not work for a manually operated digital stop watch, for which the uncertainty is about 0.2 s even though the divisions are in 0.01 s, because human reaction times will delay the start and stop signals and add uncertainty to the time interval.

For digital meters, the manufacturer gives a value of the uncertainty in the last digit; a typical value might be 0.02.

A typical school digital ammeter will read to 2 decimal places, have an uncertainty of 2% and a *resolution* of 0.02. A resolution of 0.02 means a reading of 2.50 could be 2.48 or 2.52. An uncertainty of 2% in a reading of $2.50 = 2/100 \times 2.50 = \pm 0.05$. So the true value of the current could lie between 2.43 and 2.57.

An estimated extra amount must be added to these values of uncertainty, to allow for reading errors such as parallax, calibration errors, poor visibility, rapid change in the quantity being measured, or other sources of systematic error.

Percentage uncertainty

It is the size of the uncertainty compared with the reading itself that is significant. An uncertainty of 2 mm in a measurement of 2 m is 0.1%, but 2 mm in a measurement of 20 cm is 1%, ten times more significant. Experiments should be designed to make the measured quantities as large as possible.

Overall uncertainty

- If quantities are multiplied or divided to obtain the final result, the % uncertainties should be added.
- If a quantity is raised to a power, the % uncertainty should be multiplied by that power.
- If quantities are added or subtracted, the *actual* uncertainties should be added. It follows from this that if two nearly equal quantities are subtracted to give a small difference, the % uncertainty becomes very large (pp. 43, 182). This is something else to avoid in designing experiments.

If the final result of an experiment is a measured quantity, this can be quoted with the overall uncertainty. For example:

'The average mass of a 1 N weight hanger was found to be 101.2 ± 0.5 g.'

If the end result is a concluding statement, it can be quoted as being true within the limits of the uncertainty. For example, 'The measurements show that it is 90% certain that gamma rays from a small source follow the inverse square law for radiation'.

The final report

Your final report should include:

- the aim and prediction of the investigation;
- the plan or 'method' you finally used to get the measurements;
- the measurements in a table;
- a graph and calculations based on the readings;
- an evaluation;
- a final conclusion linked to the aim.

The report should be easy to follow. Be concise; be logical; use technical terms in short sentences and make sure the spelling and punctuation are perfect!

SECTION

B Experiments and investigations

This section is divided into topics from the A level Physics course. The practical tasks within these topics are presented as *preliminary work*, *experiments*, *computer simulations* (which have the CD symbol to show you that you will need the CD accompanying this book), and *investigations*.

Preliminary work introduces the topic area and links with ideas that you may have met earlier. *Experiments* include sections on planning, analysing and evaluating, and are examples of how to structure a report on an experiment. *Computer simulations* use theoretical equations and computer graphics to model experiments that are difficult to do in a real laboratory. They allow you to control variables, take readings, plot graphs and explore the consequences of the model, but they are not true experiments and cannot lead to new discoveries.

Investigations are more open-ended experiments, which can be taken as starting points for investigations of your own. Ideas are given on how to extend them. Some of them are called *Full investigations*, and these are sufficiently extensive on their own.

The *Sample readings* given are all genuine readings taken by students or the author. The tables of readings provide examples of how results can be presented. Those with a CD symbol by them are provided as spreadsheet files on the accompanying CD. The little pens point to written tasks for you to do.

An extra, useful task would be to write a brief report on each of the experiments. The use of clear, simple language to describe what you have done is a skill worth developing.

5 Motion

5.1 Motion anaylsis

Preliminary work: Measuring average speed
Experiment: Doing a speed/time analysis of the motion of a model car
Computer simulation: Skater

Preliminary work

Measuring average speed

Apparatus

- clockwork model car, or trolley that can be accelerated by a length of stretched elastic
- track or clear bench area
- stop watch for each member of the group
- metre rule

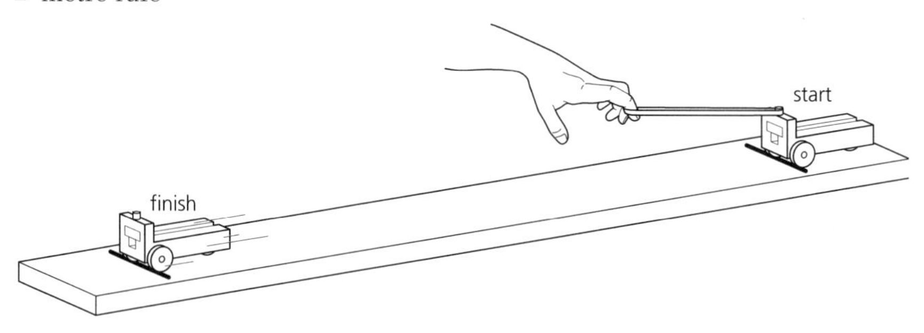

Figure 5.1

Plan

- Draw starting and finishing marks on the track 1 to 2 metres apart.
- Wind up the car or stretch the elastic and hold the vehicle ready on the starting mark.
- On a signal, release the vehicle and start the watches. Stop the watches when the car reaches the finishing mark.

Analysis

1. Record each person's time to the number of decimal places shown on the watches. Average them and round the value to 1 decimal place.
2. Calculate the average speed from distance/time.

Experiment

Doing a speed/time analysis of the motion of a model car

Apparatus

- ticker timer
- ac power supply and leads
- ticker tape
- adhesive tape
- clockwork car (or trolley that can be catapulted by elastic bands)
- track or clear bench area
- long ruler

Plan

- Adjust the power supply to the voltage required by the ticker timer.
- Connect the ticker timer to the ac sockets of the power supply. Pull a piece of ticker tape through the timer and check that it prints clear dots.
- Wind up the car and stick one end of a 1 metre length of tape to the top of it. Thread the tape through the ticker timer, turn on and release the car.
- Print a ticker tape of the motion for each member of the group.
- As an extension, you could vary the driving force to see how it alters the speed/time graph.

Analysis

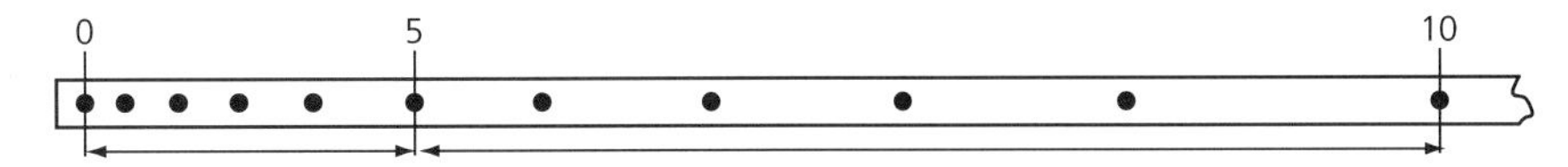

Figure 5.2

1 Draw a mark through every 5 dots on the tape, labelling the first dot zero.
2 Measure the lengths in centimetres of the 5-dot pieces. Put the distances and times into a table.

50 dots are printed every second and so the lengths of the 5-dot pieces show the distances travelled every 0.1 s. Therefore the (average) speed over a 5-dot interval will be distance/0.1 (in cm s^{-1}).

3 Calculate the speeds and plot a speed/time graph of the motion.
4 Conclude with a statement about how the speed of the car varies during the journey.

Sample readings

Clockwork car

time /s	0	0.1	0.2	0.3	0.4	0.5	0.6	0.7	0.1	0.9	1.0	1.1	1.2	1.3	1.4
distance in 0.1 s /cm	0	1.1	2.7	4.2	5.7	7.1	8.4	9.9	11.0	12.1	12.8	113.2	13.3	13.0	12.0
speed /cm s^{-1}	0	11	27	42	57	71	84	99	110	121	128	132	133	130	120

Trolley pulled by an elastic band

time /s	0	0.1	0.2	0.3	0.4	0.5	0.6
speed /cm s^{-1}	0	41	109	147	174	179	173

Plot speed/time graphs for the sample readings for the trolley and the car, and comment on the similarities and differences in their motions.

Computer simulation

Skater

Aim **To obtain matching graphs of distance/time, speed/time and acceleration/time for friction-free motion.**

Apparatus
- computer running the program 'Skater' from the CD

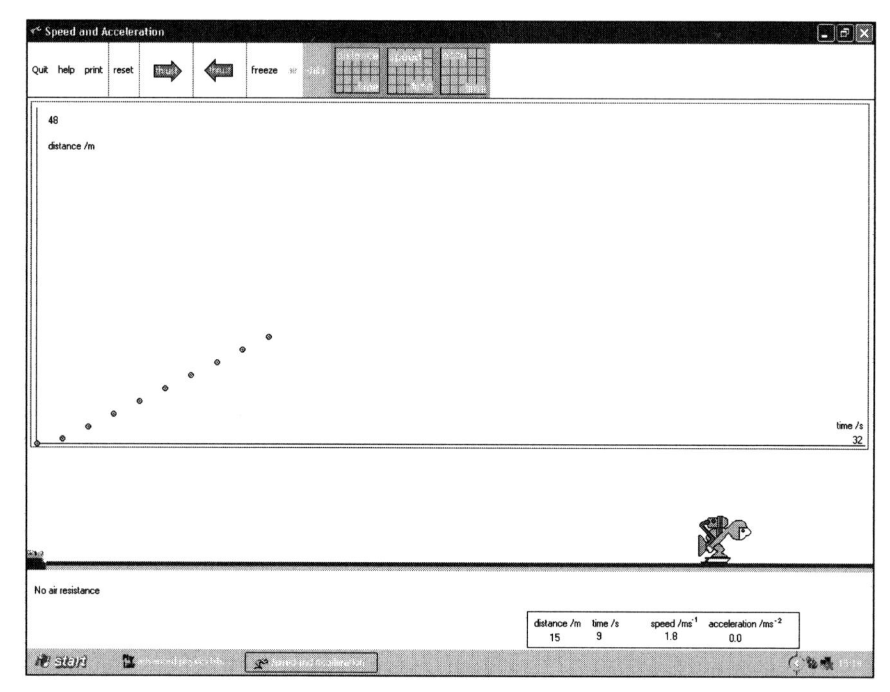

Plan The idea is to send a rocket-powered skater on a gentle journey across the ice.

- Click on forward (→) 'thrust', and the rocket engine gives the skater an acceleration. When you click it again the engine stops and she continues at a steady speed. Reverse (←) 'thrust' will slow her down and eventually move her back the other way.
- Set the skater off on a journey and try to return her gently to the start. Click on the three graph buttons to see how distance, speed and acceleration change during the motion.

Analysis

1 Print a copy of the results and draw lines through the plotted points on the graphs.

2 Add notes at key points on the separate graphs to show how they are related.

5.2 Acceleration measurements

Preliminary work: Measuring the acceleration of a trolley running down a slope
Experiment: Using a speed/time graph to measure the acceleration of a trolley down a slope
Investigation: Using a datalogger to measure acceleration

Preliminary work

Measuring the acceleration of a trolley running down a slope

Apparatus

- long runway
- stop clock or watch
- trolley
- metre rule
- buffer to prevent the trolley falling off the end of the runway

Plan

- Set the runway at a small slope by propping up one end on a pile of books.
- Make sure the wheels of the trolley are clean and turn freely. Place it on the runway and check that it accelerates when it is released.
- Mark start and finish lines on the runway 1.0 m apart.
- Measure the time it takes for the trolley to accelerate from a standing start to the finish line.
- Repeat the time measurement 5 times.

Analysis

1. Calculate the average time (t) it takes the trolley to cover the 1.0 m track.
2. Calculate the average speed $= 1.0/t$ (in m s^{-1}).
3. Calculate the acceleration:

speed at end $= 2 \times$ average speed $= 2.0/t$

(this assumes the acceleration is uniform)

acceleration = increase in speed/time taken $= 2.0/t$ divided by t

acceleration $= 2.0/t^2$ (m s^{-2})

Evaluation

This is not a precise way to obtain the value of the acceleration because of the uncertainty in the time measurement. Our reaction-time delay to the start and stop signals will introduce an uncertainty of about 0.2 s in a time of not much more than 1 second ($\approx$ 20%). This time is then squared to obtain acceleration and so the uncertainty doubles to $\sim$40%. The timing technique needs to be improved to give a more precise result for the acceleration.

Experiment: Using a speed/time graph to measure the acceleration of a trolley down a slope

Apparatus

As for the Preliminary work, plus:

- ticker timer
- ticker tape
- ac power supply set to the voltage of the ticker timer
- connecting leads

Plan

- Attach about 1 m of ticker tape to the trolley and feed it through the ticker timer.
- Switch on the ticker timer and then release the trolley so that it accelerates down the runway, pulling the tape behind it.
- Print a ticker tape for each member of the group.

Analysis

1 Mark the ticker tape every 5 dots and measure the lengths in centimetres of the 5-dot pieces.
2 Calculate the average speed for each of the pieces (= distance/0.1).
3 Put all the measurements and calculations in a table.
4 Plot a graph of speed against time.
5 Draw a best-fit straight line if it is justified by the points and calculate the acceleration from the gradient.

Sample readings

time /s	0.0	0.1	0.2	0.3	0.4	0.5	0.6	0.7	0.8	0.9	1.0
distance in 0.1 s /cm	0.0	2.7	5.3	7.2	9.7	12.0	14.0	15.9	17.6	19.9	21.7
speed /cm s^{-1}	0.0	27	53	72	97	120	140	159	176	199	217

Plot a speed/time graph for this data. From the graph calculate the acceleration of the trolley (gradient of the line) and the total distance travelled in 1 second (area under the graph).

Answers

2.25 m s^{-2}; 113 cm.

Evaluation

The times and distances measured from the ticker tape can be very precise but the drag of the tape through the ticker timer reduces the acceleration that the experiment sets out to measure. This is a design flaw that limits the accuracy of the experiment.

Investigation: Using a datalogger to measure acceleration

Apparatus

- air track and blower (or the long track from the experiment above)
- V-shaped vehicle that rides on the track (or trolley)
- light sensor connected via an interface (ADC) to a computer running a program that can record rapid changes in light intensity
- small torch or lamp to illuminate the light sensor
- piece of stiff black card about 15 cm long fitted onto the vehicle/trolley

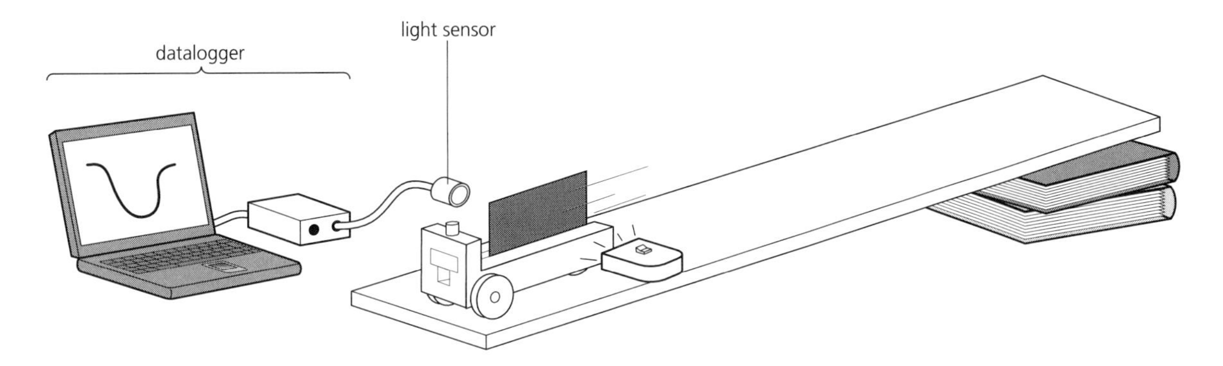

Figure 5.3

Plan

- With the air track, prop up one end with books, turn on the blower and make sure the vehicle accelerates gently down the slope. Arrange a soft buffer at the end to catch it. (With the track and trolley, set up as in the previous experiment.)
- Position the light sensor and lamp on either side of the track near the bottom of the slope. Run the computer program and select the option that samples the light level every 20 ms and shows the level on a graph.
- Fix the black card on to the vehicle and check that when the card lies between the lamp and the sensor, there is a sharp drop in the displayed light level reading.
- Measure the length (c) of the card in centimetres.
- Release the vehicle from a start mark at the top of the track and use the software to obtain the time it takes the card to cross the light beam at the bottom.
- Do 5 runs and get an average value (t_{av}) for this time.
- Measure the distance (s) from the start mark to the light sensor.

Analysis

The speed of the vehicle at the bottom of the slope $v = c/t_{av}$.
Using $v^2 = u^2 + 2as$,

$a = v^2/2s \qquad (u = 0)$

Calculate the final speed v and the acceleration a.

Sample readings

Length of black card = 17.9 cm
Distance from start line to light sensor = 114.0 cm

Readings from the computer trace of the dip in light level

start of dip /ms	100	160	120	100	190	40	95
end of dip /ms	370	420	380	370	460	295	355
time of dip /ms	270	260	260	260	270	255	260

Average time for passage of card = 262 ms
Final speed of vehicle $v = 0.179/0.262 = 0.683$ m s^{-1}
Acceleration $a = 0.683 \times 0.683/(2 \times 1.14) = 0.21$ m s^{-2}

Skill level (Analysing)

A: I put my results into a table. All columns (or rows) had headings with units. I calculated the average time of passage. I calculated the acceleration to the same number of sig. figs as the data. I wrote a concluding sentence.
All but one of the above = B; all but two = C; all but three = D; all but four = E.

Evaluation This method avoids the systematic error introduced by the drag of the ticker tape in the previous experiment. The chief source of uncertainty is in deciding exactly when the card has passed and the light is back 'on'. Because of the finite sizes of the light and sensor, the 'on' and 'off' can last 5 ms, giving an uncertainty of 10 ms in 260 ms ($\approx$ 4%). Speed is squared to get acceleration and so the overall uncertainty is at least 8%.

Extension To make this a full investigation, you could find how the acceleration of the trolley depends on the slope of the track, or on the distance it has travelled down the track.

5.3 The acceleration of an object pulled by the force of gravity

Preliminary work: Estimation
Experiment: Using a ticker timer
Investigation 1: Using automatic timing
Investigation 2: Using a datalogger

Preliminary work — Estimation

Apparatus
- stone about the size of a golf ball
- piece of thick cardboard to protect the floor

Plan
- Hold out the stone at arm's length and drop it onto the cardboard.
- Ask other members of the group to guess the time of fall.

Analysis Work out the average of the estimates of your friends.

Sample estimates

1.0 s	1.0 s	0.5 s	0.5 s	0.5 s	1.0 s	1.0 s	0.5 s

Average of estimates = 0.75 s

Evaluation Guessing the time of fall is probably no less precise than using a stop watch. The reaction time for an anticipated event might be 0.2 s, so starting the watch could be 0.2 s late and stopping it could be delayed by a similar amount. The delays might cancel to some extent but the uncertainty will remain and be at least 0.2 s in about 0.5 s ($\approx$40%). A more precise method of timing is essential.

Experiment

Using a ticker timer

Apparatus

- stone about the size of a golf ball
- 50 Hz ticker timer and power supply
- ticker tape
- adhesive tape
- metre rule
- thick cardboard (~A3 size) to protect the floor

Plan

- Attach a 2 m length of ticker tape to the stone and thread the paper through the ticker timer. Connect the timer to a suitable ac supply.
- Ask a tall person to hold the ticker timer vertically, as high as possible. Another person should hold the stone by its paper tail so that it is just below the timer. (Do not stand on stools for this activity.) Switch on the timer and release the stone.
- Examine the tape to make sure that clear single dots have been printed on it.
- Repeat 4 times with fresh paper tape. Some tapes may have to be rejected because they give anomalous results.

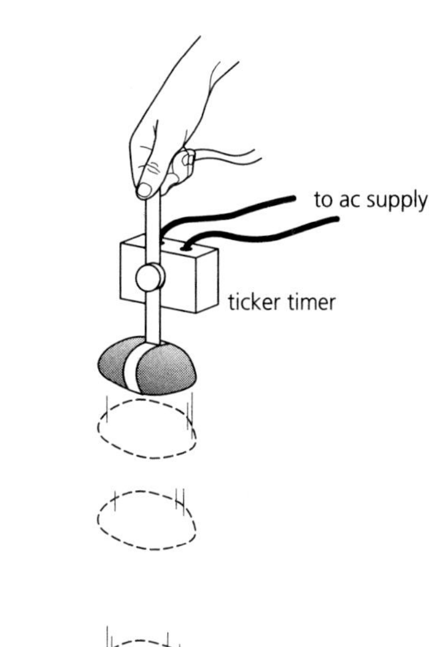

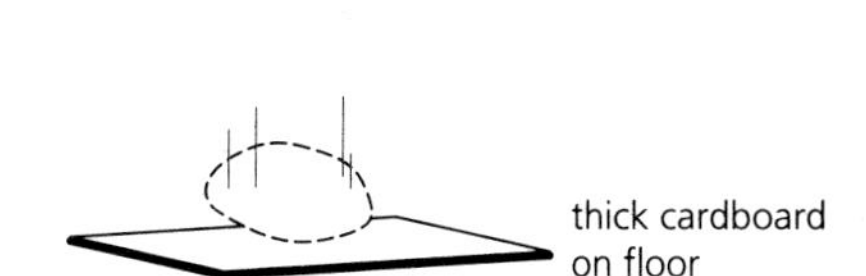

Figure 5.4

Skill level (Implementing)

A: I connected the timer to the correct supply voltage. I succeeded in printing 5 tapes with clear dots. The power supply and leads were neatly and safely arranged. No damage was caused to the floor or anyone's feet and nothing fell over.

All but one of the above = B; all but two = C; all but three = D; all but four = E.

Analysis

1. For each tape, count as many dots as you can that show acceleration (i.e. have increasing dot separation), marking the first dot zero. Measure the distance (s) covered by these dots.
2. Calculate the time of fall (t = number of dots/50).

$$s = ut + \tfrac{1}{2}at^2$$
$$s = \tfrac{1}{2}at^2 \qquad (u = 0)$$
$$a = 2s/t^2$$

3. Put the readings and results from each tape into a table and calculate the average value of the acceleration.

Sample readings

	1st tape	2nd tape	3rd tape
number of dots	30	30	30
time /s	0.60	0.60	0.60
distance /cm	164.4	165.8	168.8
acceleration /m s^{-2}	9.13	9.21	9.39

Average acceleration $= 9.2 \pm 0.1$ m s^{-1}

Evaluation If the dots have been printed cleanly on the tape, the time and distance measurements will be precise. You may have to reject some of the printed tapes because the stone did not fall cleanly and the distance measurements were wildly out. An estimate of the uncertainty can be made by comparing the values for acceleration obtained from each tape. In the sample readings the variation of the readings from the average is 0.1 in 9.2, or about 1%.

The drag of the tape as it passes through the ticker timer will reduce the acceleration of the stone and so the result is *not* the acceleration of 'free fall' of the stone.

Investigation 1 Using automatic timing

Apparatus

- 1/100th second timer that can be started and stopped by electrical contacts
- large steel ball bearing
- 2 metre rules
- start and stop gates that can automatically operate the clock (commercial kit or home-made from cardboard and aluminium foil as in Figure 5.5)
- something to catch the ball in

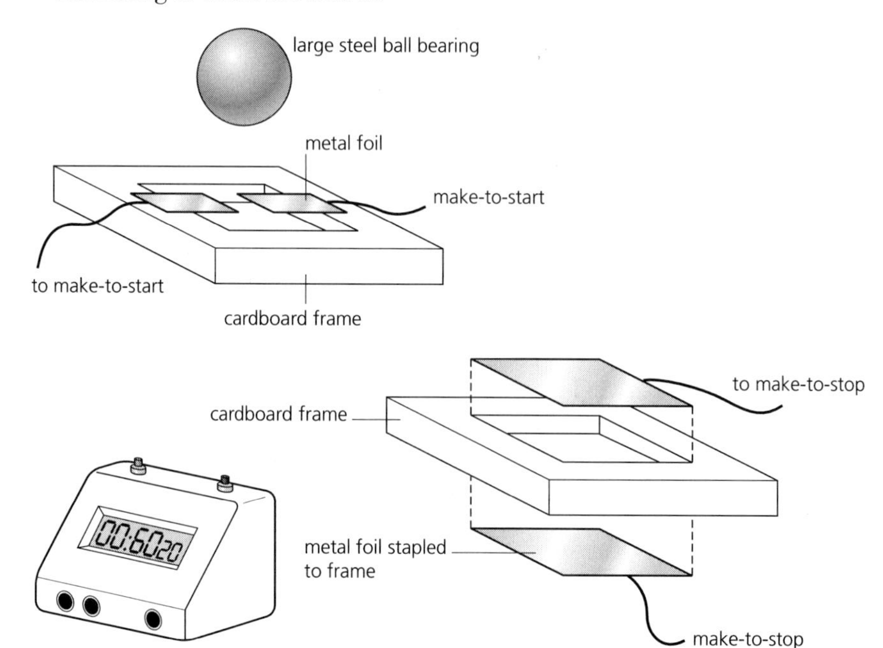

Figure 5.5

Plan

- Build the apparatus so that when the ball is released the clock starts and when the ball lands the clock stops.
- When it is working, release the ball and get at least 10 values of the time of fall.
- Measure the distance from the bottom of the ball before it is released to the bottom contact.

Skill level (Implementing)

A: I wired up the clock so that it could be started automatically. I added a switch that stopped the clock automatically. I was able to get the falling ball to start and stop the clock. I got 10 readings of the time of fall. The equipment was arranged neatly and safely and was left undamaged.

All but one of the above = B; all but two = C; all but three = D; all but four = E.

Analysis

1 Calculate the average time of fall from your results.
2 Calculate the average speed of fall (distance/average time).
3 Calculate the speed at the bottom of the drop (2 × average speed).
4 Calculate the acceleration (speed at bottom/average time).
5 Put all your readings and calculations into a neat table.

Sample readings

Distance dropped = 1.80 m

time of fall/s	0.58	0.59	0.62	0.57	0.61	0.61	0.63	0.59	0.60	0.60

From these sample readings calculate the average time of fall and the acceleration of the ball.

Answers

0.60 s; 10.0 m s^{-2}.

Evaluation

The uncertainty in the times can be estimated from the spread of the readings from the average. For the sample readings the variation from the average is about 0.02 in 0.60 (≈ 3%). The time is squared in the formula for acceleration and so the uncertainty becomes 6%.

The distance can be measured to 0.5 cm, allowing for movement that can occur when the ball and switch are reset. This will be an uncertainty of ~0.3%.

The overall uncertainty in acceleration will therefore be about 7%, due mostly to the time measurement.

There could be other systematic uncertainties that do not show up in the variation of the readings. The release of the ball from an electromagnet could be delayed, for example, or the ball could be travelling when it passes through the first gate. The inaccuracy caused by factors such as these are difficult to estimate but their presence should be noted, together with suggestions of how they might be reduced.

Investigation 2 Using a datalogger

Apparatus

- strip of black card about 15 cm long, weighted with Blu-tack (or similar)
- light sensor connected to an interface (ADC) and computer
- small torch or bulb to shine light on the sensor
- metre rule

Plan

- Arrange the sensor and torch so that the card can be dropped between them.
- Fix a marker, about 1 m above the light sensor, from which the card will be dropped.
- Program the computer so that a drop in light level triggers it to start taking readings of the light level every few microseconds.
- Drop the card a number of times from the same height until you get a good trace of the light cut-off (see the *Sample readings* below). Determine the cut-off time from the trace.
- Measure the length of the card and the distance (s) that the bottom of the card falls before the light is cut off.

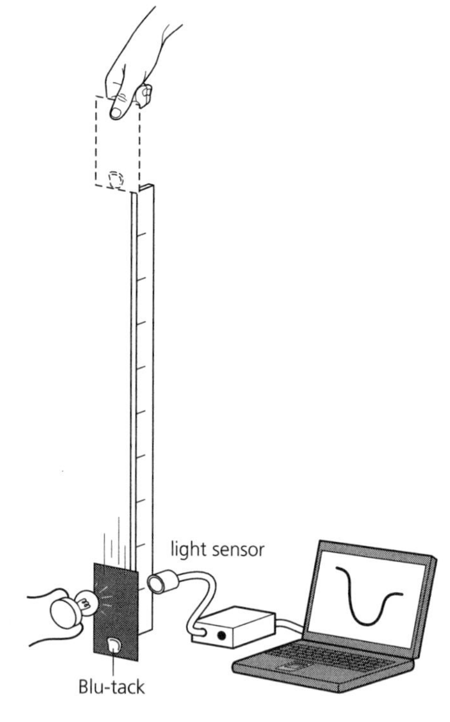

Figure 5.6

Analysis

1 Calculate the average speed (v) of the card as it passes the light sensor:

v = length of card/crossing time

$v^2 = u^2 + 2as$

$a = v^2/2s \qquad (u = 0)$

2 Use this equation to calculate the acceleration (a).

Sample readings

Length of black card = 17.8 cm

Distance fallen = 78.0 cm

The graph [opposite] was plotted from the readings taken by the ADC and computer of the drop in light level as the card passed the light sensor.

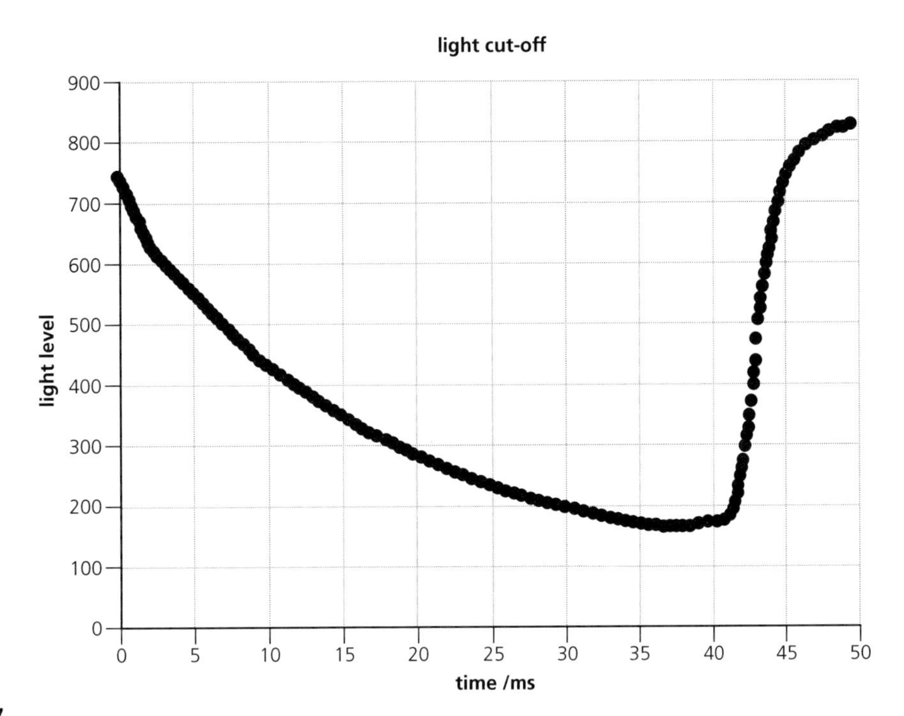

Figure 5.7

Estimate the cut-off time from this graph. Calculate the average speed of the card at the bottom of its fall. Calculate its acceleration.

Answers ~45.5 ms; ~3.91 m s^{-1}; ~9.80 m s^{-2}.

Evaluation The finite sizes of the light and the sensor can make it difficult to judge exact 'on' and 'off' times. The trigger will also add a short delay to the start of timing. The uncertainty could be a total of 5 ms in 45 ms (≈10%), leading to a 20% uncertainty in the acceleration. The length measurements are much more precise than this.

Air resistance is a source of inaccuracy that will delay the card and prevent it from 'free-falling'.

Extension To make this a full investigation, you could measure the final speed (v) for different dropping distances (s) and use the equation of motion $v^2 = 2as$ to calculate the acceleration a from a graph.

6 Force and power

6.1 Forces in equilibrium

Computer simulation 1: Identifying forces
Computer simulation 2: Triangle of forces
Experiment: Using the triangle of forces to find the weight of a Bunsen burner

Computer simulation 1 Identifying forces

Aim **To identify forces and where they act.**

Apparatus

- computer running the program 'Forces' from the CD

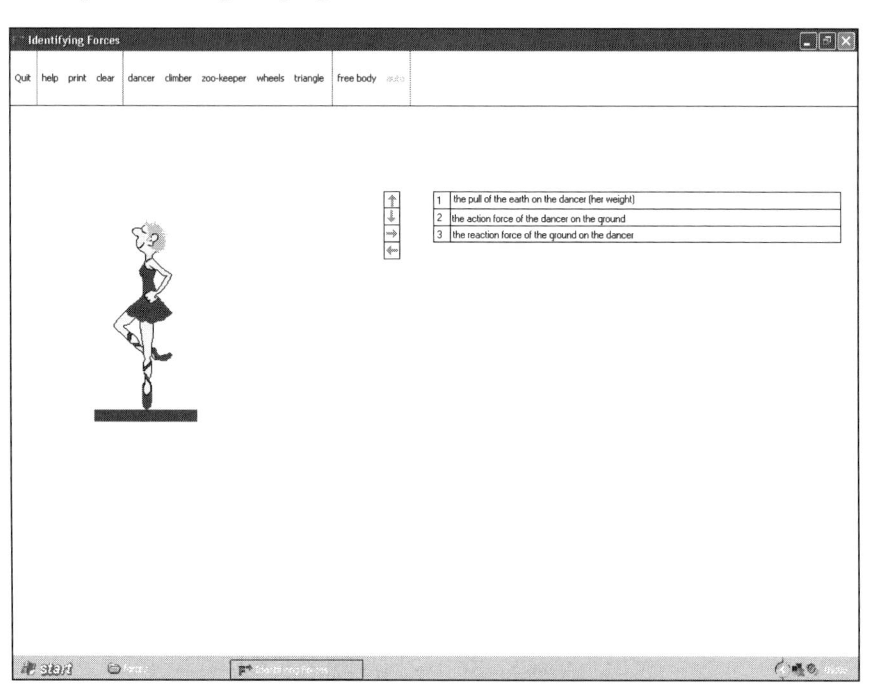

Plan

- Choose one of the scenes and click the 'free body' button. This separates the picture into different parts on which you can identify the forces acting.
- The forces are described in numbered boxes. Click on the number of the force and click on an arrow to choose its direction.
- Move the cursor to where you think the force should act. When you are in the right place the cursor will turn into a star. Click, and the force arrow will be drawn.
- Place all of the forces in this way and print out the free-body diagram for your notes.
- Do the same for the other scenes.

Computer simulation 2

Triangle of forces

Aim **To show how the vectors of three forces in equilibrium, drawn end to end, form a closed triangle.**

Apparatus
- computer running the program 'Forces' from the CD

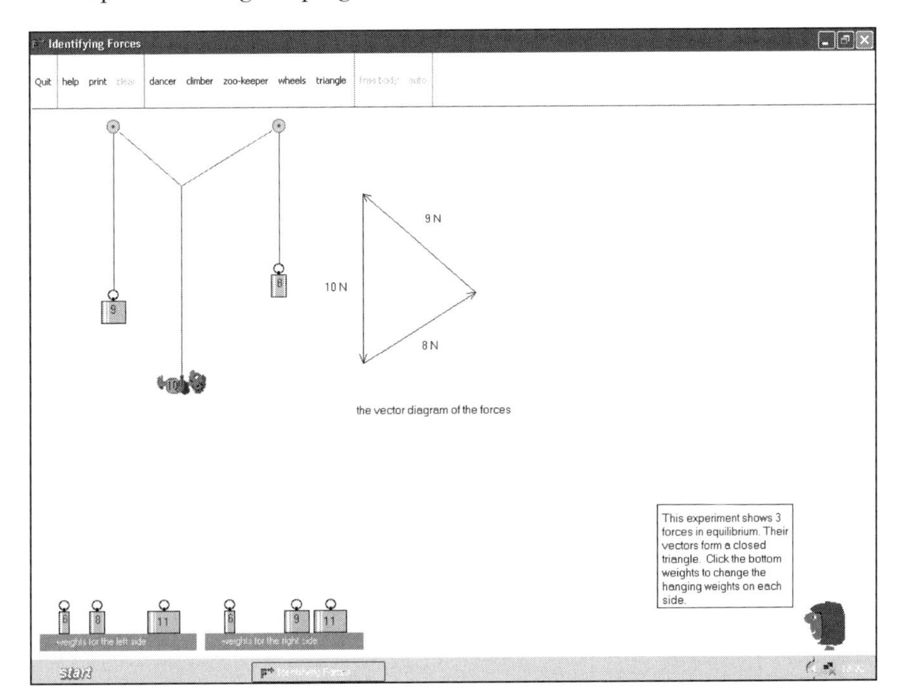

Plan
- Click 'triangle', and a picture comes up showing a 10 N baby supported by two weights on strings. The three forces that act where the strings meet are in equilibrium. The vectors that represent these forces are drawn to scale on the right of the picture. The vectors are drawn end to end and form a closed triangle.
- Change the hanging weights by clicking on the ones on the table. Watch how the angles change until equilibrium is re-established. Notice also how the triangle of forces changes.
- Print a copy of the picture for each of the possible force arrangements.

Experiment: Using the triangle of forces to find the weight of a Bunsen burner

Apparatus

- 2 pulley wheels, 2 clamps, bosses and tall stands (or 2 pulleys fixed to a board)
- strong thread
- 2 weight hangers (1 N) and extra 1 N weights
- torch
- white paper pinned to a drawing board

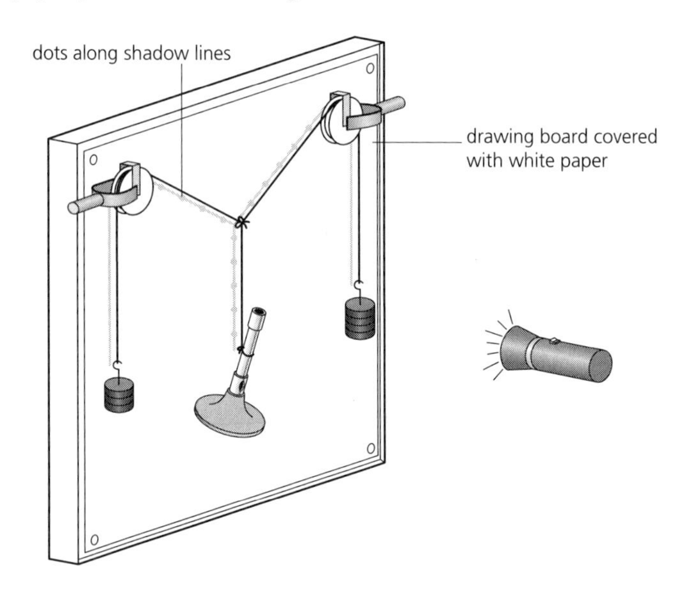

Figure 6.1

Plan

- Cut three 0.5 m lengths of thread and tie their ends together.
- Clamp a pulley at the top of each stand and place them about half a metre apart.
- Lay a piece of thread over each pulley and tie weight hangers on the ends. Tie the Bunsen burner to the third piece of thread so that it hangs between the pulleys (see Figure 6.1).
- Add weights to the hangers until they support the Bunsen.
- Hold the drawing board close to the threads and use a distant torch to cast a shadow of the threads on the paper. Dot along the shadows so that later you can draw in the shadow lines.
- Use different forces to support the Bunsen by changing the weights. Use a fresh piece of paper and record the positions of the threads as before.

Analysis

1. Draw straight lines through the dots on the paper to obtain the angles made by the threads.
2. Construct a triangle of forces for the three forces. First draw the two known forces to scale, end to end, and at the angle shown by their threads.
3. Complete the triangle with a third line. Measure the length of this line and, using the same scale, calculate the weight of the Bunsen burner.
4. If you used different weights to support the Bunsen, draw their triangle of forces and obtain extra values for the weight of the Bunsen. These can then be averaged and compared with the value calculated from the mass given by a balance.

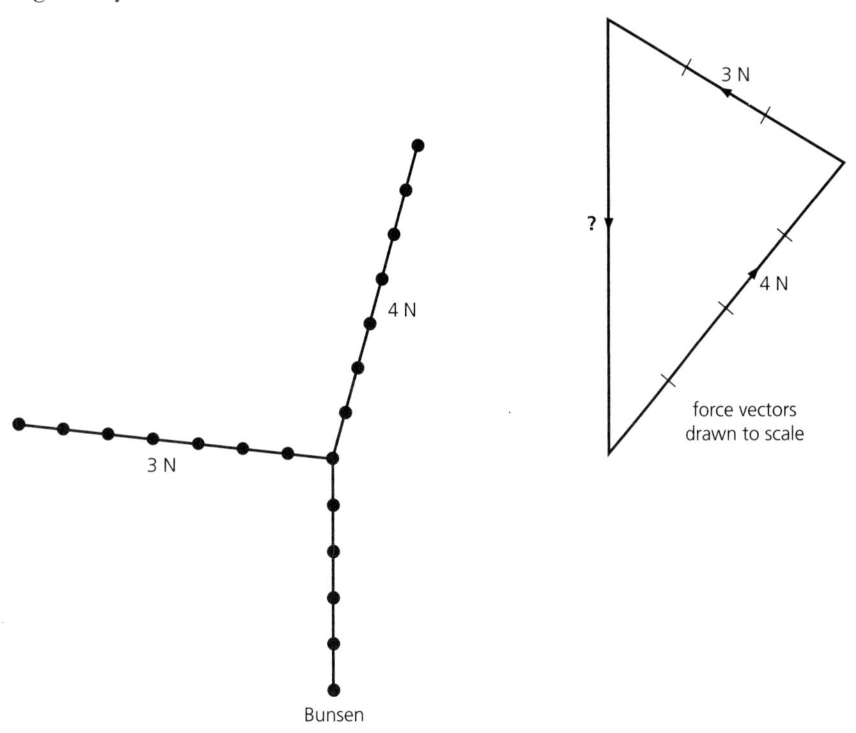

Figure 6.2

6.2 The principle of moments

Experiment: Measuring the mass of a metre rule using the principle of moments
Investigation: Measuring the density of glass

Experiment: Measuring the mass of a metre rule using the principle of moments

Apparatus

- clamp, boss and stand
- undamaged metre rule
- 0.1 N weight hanger and extra 0.1 N weights
- thread

Plan

- Cut a length of thread (about 15 cm) and tie a loop at one end that will go over the metre rule. Tie the other end to the 0.1 N weight hanger.
- Fix the clamp to the stand so that its bar is horizontal.
- Hang the weight hanger near one end of the metre rule and place the metre rule on the bar (see Figure 6.3).
- Move the rule until it balances horizontally.
- Record the positions of the weight hanger and the pivot (support point) on the rule.
- Repeat with extra weights on the hanger.

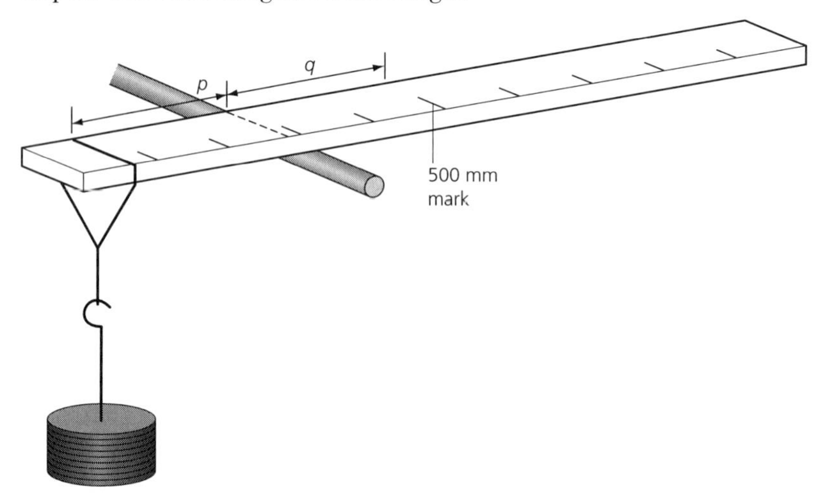

Figure 6.3

Analysis

1 Set up a spreadsheet to record the mass H of the hanger, its position on the rule and the position of the pivot. You will need to enter formulae to calculate the distance from the hanger to the pivot (p) and the distance of the centre of mass of the rule (assumed to be at the 500 mm mark) to the pivot (q) (see point **2**).

The moment of the hanger about the pivot = Hgp
The moment of the mass (W) of the rule = Wgq

When the rule is balanced:

$$Hgp = Wgq$$
$$W = Hp/q$$

2 Enter formulae in the spreadsheet to do all of the calculations for you and to get an average result for the mass of the rule:

Put the measured quantities in the top three rows of the spreadsheet. Enter the following formulae in the next three rows and 'fill across' (go to Edit–Fill–Right) to column G. (See the *Sample spreadsheet* below.)
In C4 put **=C2–C3** (this works out the distance *p*). Include the '='.
In C5 put **=C3–500** (this works out the distance *q*).
In C6 put **=C1*C4/C5** (this is the formula to calculate the mass of the rule).
To work out the average in H6 put **=AVERAGE(C6:G6)**.
To work out the standard deviation in H7 put **=STDEVP(C6:G6)**.
You should check some of the calculations by hand.

Sample spreadsheet  The Excel file for this spreadsheet of sample results is on the CD.

	A	B	C	D	E	F	G	H	I
1	hanger mass/g		100	80	70	50	10		
2	position of hanger/mm		990	990	990	990	990		
3	position of pivot/mm		755	725	710	671	547		
4	distance p/mm		235	265	280	319	443		
5	distance q/mm		255	225	210	171	47		
6	mass of rule/g		92.2	94.2	93.3	93.3	94.3	93.4	average
7								0.8	std

Evaluation The chief uncertainty arises from getting the beam to balance horizontally and reading the position of the pivot. The position has to be found by trial and error, and read from a scale that lies above the actual point of contact. The uncertainties in each of the measured distances will be about 2 mm in 20 cm (=1%). There will also be a small uncertainty in the weights of the hanger and masses. The overall uncertainty in the value for the mass of the ruler could be as little as 2%. (Check against the value given by an electronic balance to see if you are within these limits.)

Improving the plan

- Measure the masses of the balancing weights accurately on an electronic balance.

Investigation Measuring the density of glass

Apparatus

- small block of glass, e.g. glass prism or glass stopper
- beaker containing enough water to submerge the glass
- undamaged metre rule
- clamp, boss and stand
- thread
- 1 N weight hanger
- 0.1 N weight hanger and 0.1 N weights (see *Improving the plan* opposite)

Plan

- Tie the weight hanger and the glass prism to 15 cm lengths of thread. Tie loops in the other ends of the thread, large enough to slip over the metre rule.
- Fix the bar of the clamp 20 cm above the bench and place the 50 cm mark of the metre rule on the bar.
- Hang the prism on one side of the rule and the weight hanger on the other. Adjust their positions until the rule is balanced horizontally when supported at its 50 cm mark.
- Measure the distance of each thread from the central balancing point.
- Submerge the glass prism in the beaker of water and adjust the threads until the rule is balanced again. The glass should not touch the sides of the beaker and should be fully covered by the water.
- Measure the distance of each thread from the balancing point again.

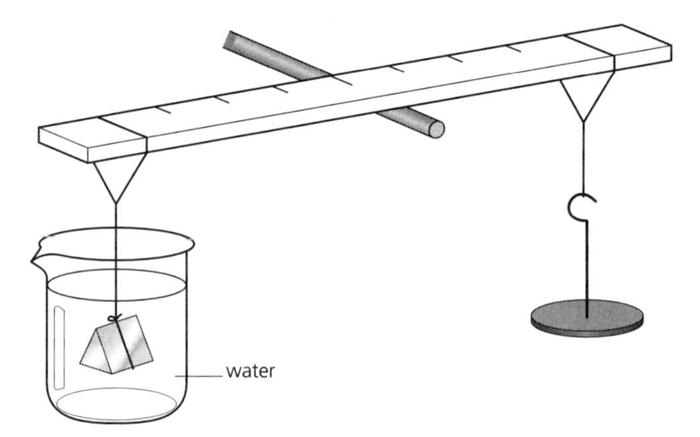

Figure 6.4

Skill level (Implementing)

A: I checked that the metre rule balanced horizontally at the 50 cm mark. I was able to balance the weight and the glass in air and in water. I placed the loops of thread as near to the ends of the rule as possible. I read and recorded the positions of the threads. The prism was fully submerged in the water and did not touch the beaker.

All but one of the above = B; all but two = C; all but three = D; all but four = E.

Analysis

1 Record your results with full descriptions and units. You should design a table for them.

2 Use the principle of moments to calculate the mass of the glass prism in air and its apparent mass in the water.

The difference between the mass in air and the apparent mass in water is the mass of water that was displaced by the glass. (This reasoning follows from Archimedes' principle.) Taking the density of water to be 1.0 g cm^{-3}, the volume of this displaced water (and hence the volume of the glass prism) in cm^3 is the same as the mass of displaced water in grams.

3 Calculate the volume of the glass from (mass in air − apparent mass in water) and then its density (= mass/volume).

Sample readings

Using a hanger of mass 100 g

	with the glass in air	with the glass in water
distance of weight hanger to pivot /cm	17.9	9.8
distance of glass prism to pivot /cm	47.5	45.0
mass of glass prism /g	37.7	21.8
mass of water displaced /g	15.9	
volume of water displaced (= volume of glass prism) /cm^3	15.9	
density of glass /g cm^{-3}	2.37	

Evaluation It is possible to read the positions of the threads with precision, but moving them to balance the rule and to find those positions is more difficult. The uncertainty in the distances to the pivot could be 2 mm in 10 cm (= 2%). When the distances are divided to calculate mass the uncertainties will add to become 4%. This gives 37.7 ± 1.5 and 21.8 ± 1 for the masses in air and in water from the sample readings.

A large increase in percentage uncertainty occurs when the two masses are subtracted. This is because the actual uncertainties are added and become a larger proportion of a smaller quantity. For the sample readings:

$$37.7 \pm 1.5 - 21.8 \pm 1 = 15.9 \pm 2.5 \ (\sim 16\%)$$

If possible, experiments should not include this procedure in their design.

Improving the plan

- Use a smaller hanging weight than 1 N to maximise the distance of the prism from the pivot. This will reduce the % uncertainty in its position.

Extension To make this a full investigation, you could measure the density of glass by another method (see p. 56) and compare the two results.

6.3 Force and acceleration

Computer simulation 1: Newton
Experiment: Measuring inertial mass using Newton's second law
Computer simulation 2: Bump

Computer simulation 1 Newton

Aim **To investigate the relationship between force, mass and acceleration.**

Apparatus ■ computer running the program 'Newton' from the CD

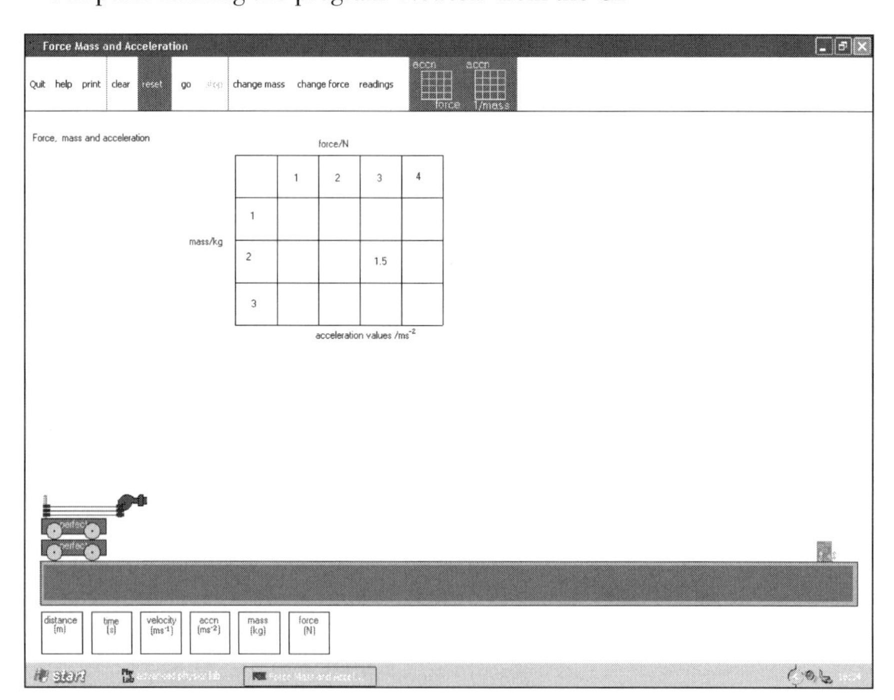

Plan The friction-free trolley in this simulation is accelerated by a length of elastic that is stretched to a constant length. When you stop the trolley, its acceleration is calculated and recorded.

- Click on 'go', and the trolley is accelerated by the stretched elastic. (For simplicity this exerts a force of 1 N.) Click 'stop' before the trolley hits the wall.
- Click on 'change force' to increase the force to '2 elastics' (this force is 2 N). Click 'go' then 'stop' to record the acceleration.
- Continue in this way to see the relationship between force and acceleration when the mass is constant.
- Next, keep the force constant and click on 'change mass'. The mass is increased uniformly by piling equal trolleys on top of one another. Investigate how acceleration changes with mass when the force is kept constant.
- Use all the possible combinations of force and mass to complete the boxes in the 'readings' grid.

Analysis

1 Print out a copy of the 'readings' grid, together with plots of the data and a selection of the 'help' pages.

2 Write a conclusion based on these results.

Experiment

Measuring inertial mass using Newton's second law

Apparatus

- runway or laboratory table that can be tilted
- trolley
- ticker timer and ticker tape
- adhesive tape
- ac power supply and leads
- 1 N weight hanger
- 1 N weights (4)
- strong thread
- pulley clamped to the end of the table

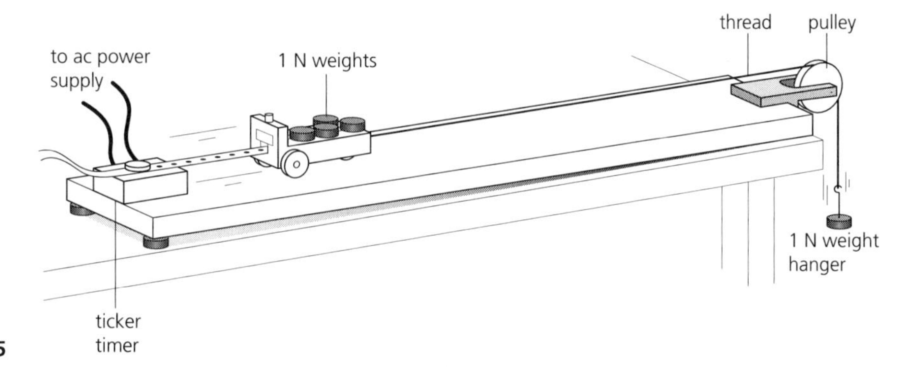

Figure 6.5

Plan

- Connect the ticker timer to the power supply that has been set to the correct voltage.
- Stick a 1 m length of ticker tape to the trolley and feed it through the ticker timer. Position the timer so that the tape runs freely when the trolley moves.
- Put the four weights on the trolley.
- Tilt the table or track until a gentle push makes the trolley run at a constant speed. Have the ticker timer running and print a tape to check that the dots are evenly spaced. (This is to compensate for friction.)
- Tie one end of a long length of thread to the trolley, pass the other end over the pulley and tie it to the weight hanger. The hanger should be as close to the pulley as possible.
- Think about safety.
- Hold the trolley at the raised end of the table, turn on the ticker timer and let go. The weight hanger will accelerate the trolley, and the ticker timer will print dots on the tape.
- Increase the pulling force by transferring a weight from the top of the trolley to the hanger. (Note that the falling weights are also accelerated and this procedure is used to keep the accelerated mass constant.) Obtain an acceleration ticker tape as before.
- Continue for all of the weights.

Skill level (Implementing)

A: I compensated for friction. I arranged the weights to fall as far as possible. I obtained ticker tapes for 5 accelerations. I had someone catch the trolley at the end, no one was injured and nothing was damaged. I left the equipment set up while I calculated the results in case a repeat reading was needed.

All but one of the above = B; all but two = C; all but three = D; all but four = E.

Analysis

1 Draw up a table (see the *Sample readings* below) or prepare a spreadsheet for your results (see p. 15).

2 Mark off 25 or 30 dots from the start of the motion that show the trolley accelerating and calculate the time (t) it took to print these dots. (50 dots are printed per second so t = number of dots/50.) Measure the distance covered (s) by this number of dots.

$$s = ut + \tfrac{1}{2}at^2$$

$$\text{acceleration } a = 2s/t^2 \quad (u = 0)$$

If F is the accelerating force, the moving mass $m = F/a$.

3 Calculate the acceleration and then the value of the moving mass.

4 Do the same for the other four tapes and calculate an average value for the moving mass. This will be the mass of trolley plus the masses of the five 1 N weights used. Subtract these from your result to get the mass of the trolley.

5 Put your results and calculated quantities into your table. Write a final result and conclusion from your work.

Sample readings

force /N	1.0	2.0	3.0	4.0	5.0		
no. of dots	40	40	40	25	25		
time /s	0.80	0.80	0.80	0.50	0.50		
distance /cm	25.8	49.6	65.0	38.0	46.4		
acceleration /cm s^{-2}	81	155	203	304	371		
mass /kg	1.24	1.29	1.48*	1.32	1.35	average mass /kg	1.30
						std /kg	0.04

*Anomalous result that should be re-measured – not included in the average.

Mass of trolley = 1.30 − 0.50 = 0.80 kg

Plot a graph of force against acceleration for the sample data, draw the best-fit line and obtain a value for the moving mass from the gradient.

Answer 1.35 to 1.38 kg.

Evaluation The uncertainty in the value for the mass of the trolley can be estimated from the spread of the values from the average, or the standard deviation. For the sample readings above this is about 0.04 kg (~3%). Time measurements from the tape will be quite precise, but the positions where the dots are printed are likely to be less certain. The weights will be accurate to 0.2% and so will not introduce much extra uncertainty into the value of the accelerating force.

Some of the force of the falling weight will be used to overcome friction and, even though the track has been tilted to compensate for this, an unknown and variable amount of friction will remain.

Improving the plan

- Use a linear air track. The vehicle is suspended on a cushion of air and there is very little opposition to motion.
- Use a light gate and datalogger to measure the final speed and acceleration of the trolley.

These steps will remove some of the systematic errors and give a more accurate result for the mass of the vehicle or trolley.

Computer simulation 2

Bump

Aim **To investigate the changes in momentum and kinetic energy when two bodies collide.**

Apparatus

- computer running the program 'Bump' from the CD

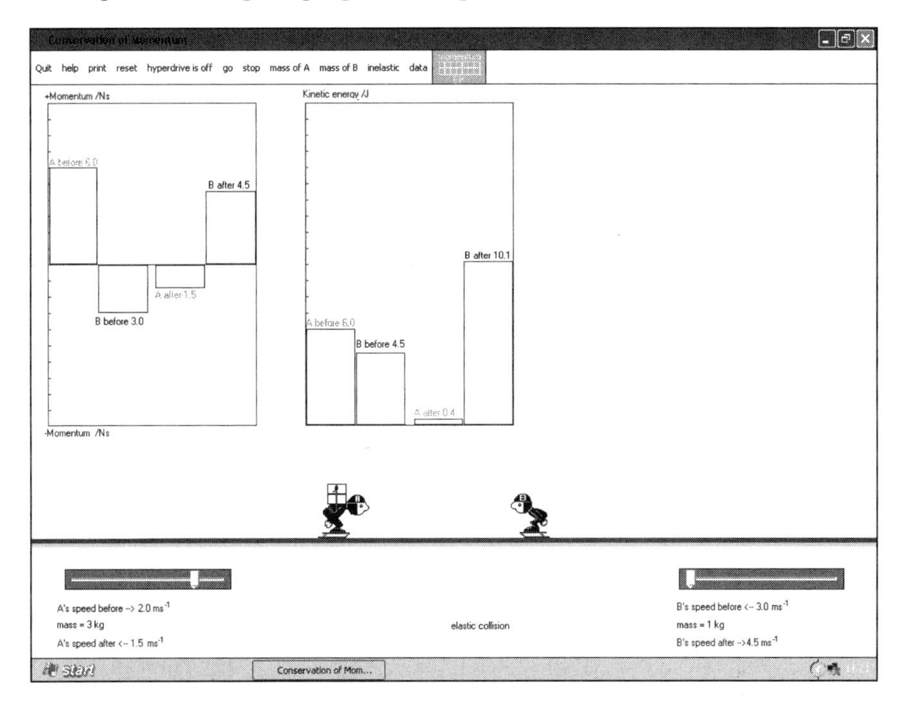

Plan

- Choose starting speeds and masses for the two skaters.
- Choose 'elastic collision' and click either 'data' or the 'momentum/Ek' button giving the bar charts.
- Click 'go' and watch the collision between the two skaters. Note how the speed, momentum and kinetic energy of each skater change.
- Repeat for a range of starting speeds and masses.
- Print out the readings and bar charts, and check whether momentum and kinetic energy are conserved in every case. Identify when the greatest exchange of kinetic energy takes place.
- Do the same with inelastic collisions.

6.4 Work and efficiency

Preliminary work: Building a pulley that can lift a load
Experiment: Measuring the efficiency of a pulley
Full investigation: Does the efficiency of a pulley vary with the load it is lifting?

Preliminary work

Building a pulley that can lift a load

Apparatus

- 2 pulley blocks with 3 wheels or 2 wheels each
- clamp, boss and stand
- G-clamp
- 3 m of thin string
- 1 N weight hangers (2)
- 1 N weights (4)

Plan

- Fix one pulley block in the clamp as far above the bench as possible. Make sure the block is held securely and that its wheels can rotate freely. 'G-clamp' the stand to the bench.
- Tie the end of the string to the hook under this pulley block.
- Thread the string down through a wheel of the other pulley block and then up through a wheel of the fixed block. Continue until the string has been passed through every wheel and the end is hanging down. Do not try to put the string in the grooves of the wheels yet.
- Hang a weight hanger with four 1 N weights on the lower block, and another weight hanger on the free end of the string (see Figure 6.6).
- Think about the consequences of the stand falling or the string breaking and take precautions.
- Finally, with the string in tension, lift it into the grooves of each pulley wheel. Make final adjustments so that the weight on the end of the string smoothly and steadily lifts the load on the lower block.
- Observe how, with this example of a useful machine, a small weight is able to lift a load of a larger weight by moving further.

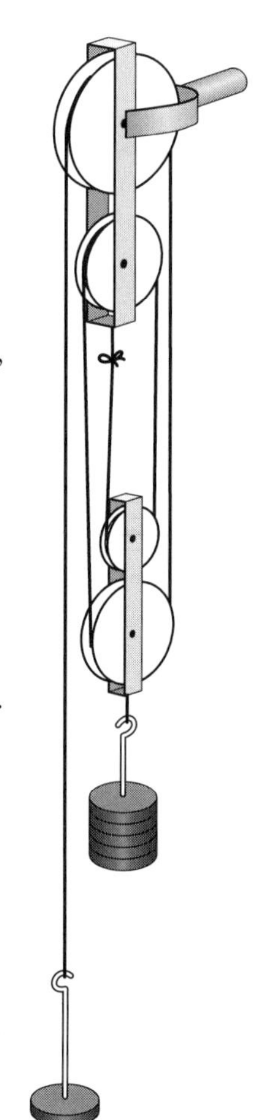

Figure 6.6

Experiment: Measuring the efficiency of a pulley

Apparatus As for the Preliminary work, plus:

- 2 metre rules

Plan

- Build the pulley (see the Preliminary work).
- Adjust the weights so that the 'effort' (smaller weight) is just able to keep the 'load' (larger weight) moving upwards at a steady speed. Give the hanger providing the effort a small push to overcome static friction and start it moving.
- Record the size of the load and the effort.
- Arrange the metre rules so that the 'effort' hanger moves down exactly 1 m. At the same time, measure the distance the load moves upwards. Record these two distances.

Analysis **1** Organise your readings and calculations into a table:

	Effort	Load
Size of the force /N		
Distance moved /m		
Work done /J		
Efficiency /%		

The falling 'effort' puts energy into the machine. The pull of gravity does work moving the hanger downwards. The 'load' gains gravitational potential energy equal to the work done on it.

Work done = force × distance moved

$$\text{Efficiency of energy conversion} = \frac{\text{work got out}}{\text{work put in}}$$

2 Calculate the efficiency of your pulley.

Evaluation There will be uncertainty in the value of the effort needed to lift the load. You have to judge when there is just enough force to lift the load but not enough to accelerate it. The uncertainty could be 0.1 N out of 4 N ($\approx$ 3%). Measuring the distance lifted for a fall of 1 m could be uncertain to 1 cm in 20 cm (= 5%), giving an overall uncertainty in the efficiency of 8%.

Full investigation — Does the efficiency of a pulley vary with the load it is lifting?

Apparatus As for the previous experiments, plus:

- 0.1 N weights

Plan

- Use the method of the previous experiment to measure the efficiency of the pulley for a small load such as 1 N. You will need small weights to get the effort just right.
- Increase the load in 1 N steps, taking readings of the effort needed to lift the load at a steady speed each time. You do not need to measure the distances lifted more than once. (The effort will always move down x times more than the load goes up, where x is the number of strings supporting the bottom block.)

Analysis

1 One way of presenting your results is in a table like this:

Load /N	1	2	3	4	5
Distance moved by load /m					
Work done on load /J					
Effort /N					
Distance moved by effort /m					
Work put into machine /J					
Efficiency /%					

2 Plot a graph of efficiency (y-axis) versus load (x-axis).

3 From your graph, write a conclusion about how the efficiency of a pulley system varies with the load it is lifting. Suggest a reason why the efficiency varies as it does.

Sample readings

load /N	1	2	3	4	5	6	7	8	9	10	11	12	13	14	15	16	17	18	19
effort /N	0.6	0.8	1.1	1.3	1.5	1.8	2	2.2	2.4	2.7	2.9	3.1	3.4	3.6	3.8	4	4.2	4.4	4.6
efficiency /%*	28	42	45	51	56	56	58	61	63	62	63	65	64	65	66	67	67	68	69

*There were 6 strings holding up the bottom block, so if the load went up 1 m the effort would have to fall 6 m.

Work got out = load × 1

Work put in = effort × 6

Efficiency = load/(effort × 6) x 100%

Plot a graph of efficiency against load for these sample results. Give a reason why the efficiency is low when the load is small.

Answer When small, the load is of similar weight to the bottom wheels of the pulley which also have to be lifted by the effort.

Skill level (Analysing)

A: I put my results into a table. All columns (or rows) had headings with units. I calculated the efficiency for each load. I plotted a properly labelled graph of efficiency against load. I wrote a conclusion including reasons for the shape of the efficiency graph.

All but one of the above = B; all but two = C; all but three = D; all but four = E.

Extension Another full investigation could be to compare the efficiency of a pulley system with that of an inclined plane and a wheel and axle machine.

6.5 Power and efficiency

Experiment: Measuring the output power of an electric motor
Full investigation: How does the efficiency of a motor vary for different loads?
Computer simulation: Work

Experiment: Measuring the output power of an electric motor

Apparatus

- pulley, clamp and stand
- weight hanger and 1 N weights
- strong thread
- metre rule
- stop clock or watch
- dc low-voltage electric motor
- dc power supply
- switch

Plan

- Connect the motor and switch to the dc supply.
- Set the voltage to about two-thirds of the motor maximum.
- Switch on and check that the motor runs.
- Build a pulley system that lifts the load by about a metre. Use thread to connect the weight hanger to the spindle of the motor. Spend some time getting the machine to run smoothly so that you can raise and lower the load by operating the switch.
- Use the stop clock to measure the time it takes to lift the load through a measured distance. Take 5 time measurements for the same distance.

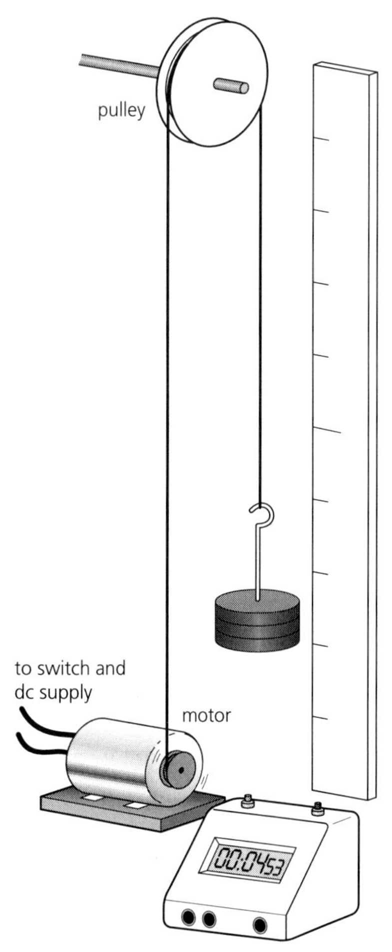

Figure 6.7

Skill level (Implementing)

A: I connected the motor and a switch to the correct voltage and made the motor run. I arranged a pulley and thread so that the motor lifted a weight in a controlled way. I fixed a mark to a metre rule and measured the time it took to lift a weight through the fixed distance. I asked a friend to help me take the readings. I got 5 values of the time for the same load and distance.

All but one of the above = B; all but two = C; all but three = D; all but four = E.

Analysis

1 Lay out your results and calculations neatly.

Work done = force × distance lifted

$$\text{Output power of the motor} = \frac{\text{work done}}{\text{time taken}}$$

2 Calculate:
- the average lift time;
- the work done in one lift;
- the output power of the motor.

Sample readings

lifting distance /m	1.0
load lifted /N	3.0
time to lift /s	4.5 4.9 5.1 4.8 5.0
average time /s	4.9 ± 0.2
work done /J	3.0
power /W	3.0/4.9 = 0.6

Evaluation

The variation of the time readings from the average is a guide to their uncertainty. The largest difference is 0.4 s and averaging reduces the likely uncertainty to 0.2 s (~4%).

The uncertainty in the lifting distance will be about 1 cm in 100 cm (= 1%) because it is difficult to stop the load at an exact place.

The overall uncertainty in the power will therefore be about 5%.

Full investigation How does the efficiency of a motor vary for different loads?

Apparatus As for the previous experiment, plus:

- voltmeter and ammeter of suitable range for the motor
- extra weights
- an assistant

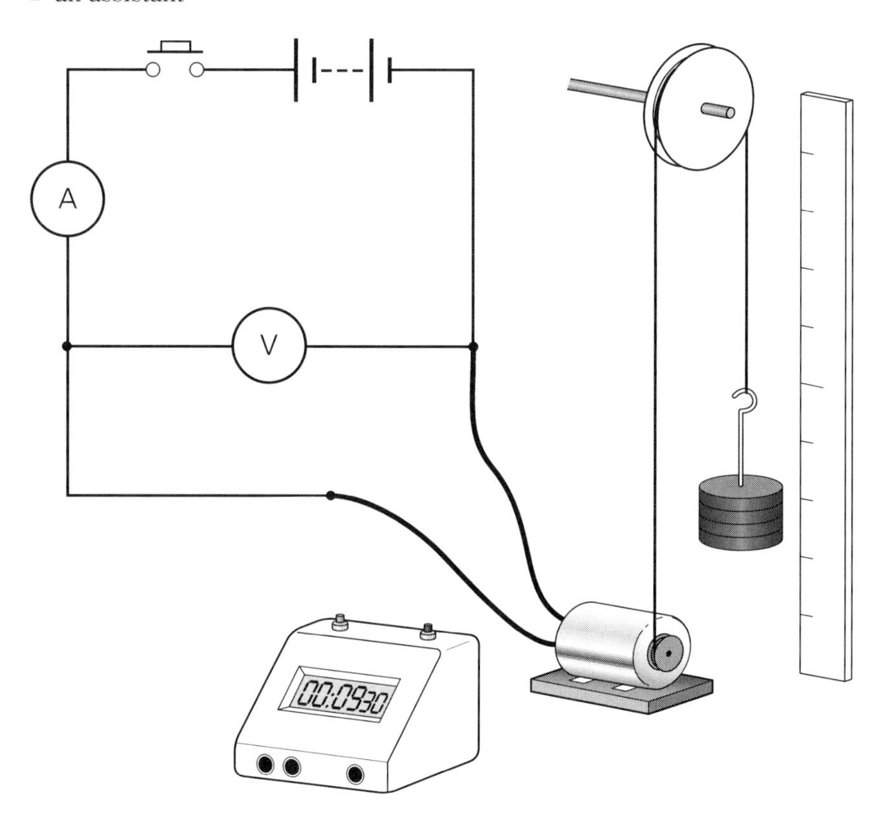

Figure 6.8

Plan

- Connect the voltmeter and the ammeter into the motor circuit and check that they record its pd and current.
- Make the motor lift a load and, as it does so, read the pd and current.
- Time the lift over a measured distance.
- Change the load systematically and repeat all the measurements.

Analysis

1 Put your results into a single table with suitable headings. Leave columns for quantities that you have to calculate (see the *Sample readings* overleaf).

Power input to motor = pd × current

Enter all measured quantities to the number of decimal places given by the instruments.

2 Plot a graph of the efficiency of the motor for different loads.

3 Write a conclusion, describing how the efficiency of the motor varies and giving your theory of why this is.

Sample readings

From the computer simulation 'Work' on the CD

force /N	distance /m	time /s	power out /W	pd /V	current /A	power in /W	efficiency /%
1.0	1.00	8.6	0.12	6.0	0.12	0.72	16
2.0	1.01	12.9	0.16	6.0	0.14	0.84	19
3.0	1.01	15.1	0.20	6.0	0.15	0.90	22
4.0	1.01	18.5	0.22	6.0	0.17	1.02	21
5.0	1.00	23.2	0.22	6.0	0.19	1.14	19

Plot a graph of efficiency for different loads, using the sample data.

Evaluation

The % uncertainty in the value of the output power will be the same as in the previous experiment (5%).

The uncertainty in the values of current and pd will be about 2% each, including changes and fluctuations as the motor is lifting.

The likely uncertainty in the efficiency will therefore be about 9%.

Improving the plan

A better method would be for the motor to work continuously against a friction belt, rather like static bicycles in a training gym. The current, pd and force values would then be more constant and timing could be done over a longer period.

Computer simulation Work

Aim **To measure the input and output power of an electric motor for different loads.**

Apparatus

- computer running the program 'Work' from the CD

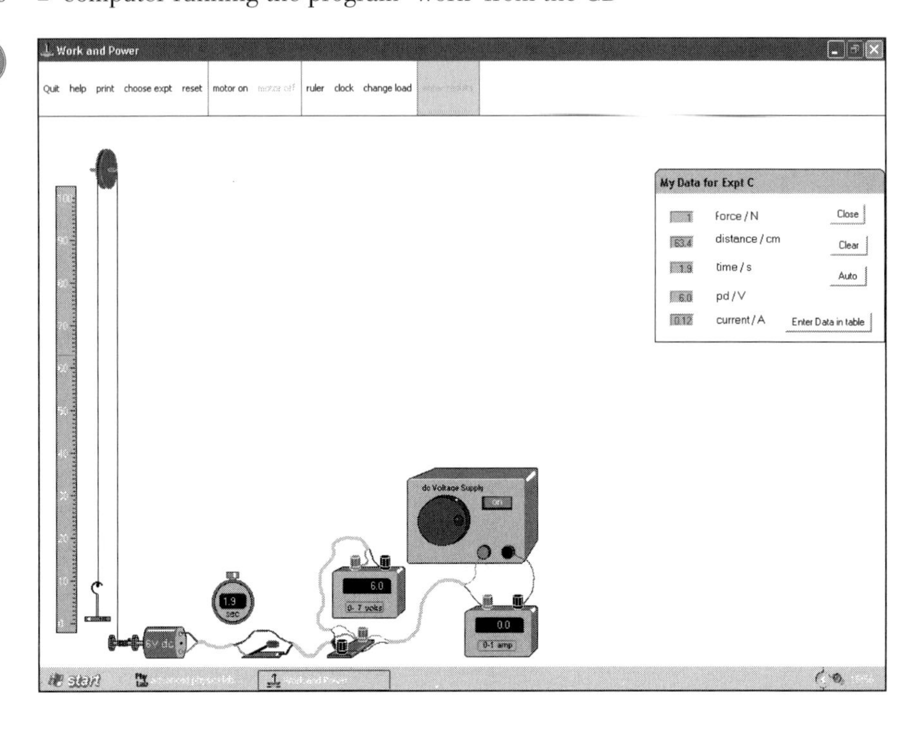

Plan The screen shows a motor, connected to a power supply and arranged to lift a load.

- Choose Experiment C.
- Switch on the power supply to the motor, close the press switch in the circuit and watch the motor lift the load.
- Wait while the motor lifts the load through a height of your choice, then switch the motor off. Record the distance, time, pd and current in the data box. The 'Auto' button will automatically record the measurements for each run. They can then be entered into the main results table, by clicking 'Enter data in table'.
- Repeat for different loads. For each set of data the input power, output power and motor efficiency are calculated.

Analysis

1 Print a copy of your results.

2 Write a conclusion about the efficiency of the motor, estimating the optimum load for maximum efficiency.

7 Materials

7.1 Density

Preliminary work: Measuring the density of solids
Experiment: Measuring the density of liquids

Preliminary work

Measuring the density of solids

Apparatus

- rectangular blocks of various materials
- selection of irregular solids (e.g. pebble, cork, candle, keys)
- overflow can
- measuring cylinders of different sizes
- ruler
- electronic balance

Plan

- Devise a way of measuring the volume of the solid by using the ruler, the overflow can and/or one of the measuring cylinders.
- Use the electronic balance to measure its mass.

See p. 42 for a method of measuring density by displacing water and measuring upthrust.

Analysis

1 Make up one large grid for the measurements of all the solids. Leave columns for the volume, density and 'density order'.
2 Calculate the density for each solid. Take care with the units and give the final density values in kg m^{-3} to the correct number of significant figures.

Experiment

Measuring the density of liquids

Apparatus

- selection of liquids, e.g. cooking oil, glycerine, paraffin
- 100 cm^3 distilled water
- density bottle
- electronic balance
- thermometer

Plan

- Make sure the density bottle is dry and clean, then use the balance to find its mass.
- Fill the bottle to its rim with distilled water and push in the stopper so that excess water squirts out through the hole in the lid.
- Dry the outside of the bottle and find its mass when full of water.
- Empty the bottle, dry the inside if possible (think about how to do this), and refill with a different liquid.
- Dry away excess liquid and find the mass of the bottle when filled with the new liquid.

Skill level (Implementing)

A: I checked that the balance was zeroed, weighed the dry bottle and recorded its mass. I completely filled the bottle with water. I dried the outside and weighed it again. I emptied the water and dried the inside. I filled the bottle with liquid, dried the outside and weighed it. I noted the temperature of the liquids.

All but one of the above = B; all but two = C; all but three = D; all but four = E.

Analysis

1 Put your readings in a results table similar to that in the *Sample readings* below.

2 Use the method to measure the densities of a number of liquids and conclude by putting them in 'density order'.

Sample readings

For glycerol at 25 °C

mass of bottle empty /g	41.28
mass of bottle full of water /g	91.64
temperature of the water /°C	25.0
density of water at 25.0 °C /g cm^{-3}	0.997
mass of water /g	*
volume of bottle /cm^3	*
mass of bottle full of liquid /g	104.36
mass of liquid /g	*
density of liquid /kg m^{-3}	*

*Quantities to be calculated from the measurements.

Evaluation

This old-fashioned experiment can be precise and accurate. It can give results for density to 4 significant figures. The uncertainty in the three mass measurements could be as small as 0.02 g. Subtracting to find the masses of the liquids increases the uncertainty to about 0.04 g in 50 g ($\approx$ 0.1%).

The chief source of uncertainty is the presence of water or air bubbles when the bottle is full of the test liquid. Small temperature changes during the experiment can also alter the actual amount of liquid in the bottle.

Extension

You could design a further experiment to measure the density of a liquid at different temperatures. If you put the full bottle into a hot water bath, the liquid will expand and some will escape through the hole in the stopper. You could then dry the outside of the bottle and use the balance to find the mass and density of the liquid at the temperature of the water bath.

Sample readings

Mass of bottle full of glycerol at 100 °C = 102.01 g

Using the sample readings, calculate the density of glycerol at 25 °C and at 100 °C.

Answers

1256 kg m^{-3}; 1210 kg m^{-3}.

7.2 Stretching springs

Preliminary work: Plotting a force/extension graph for a spring
Experiment: Measuring the elastic limit and spring constant of a spring
Investigation: Measuring the spring constant of springs in parallel and in series

Preliminary work — Plotting a force/extension graph for a spring

Apparatus

- laboratory spring (e.g. one for which 1 N gives an extension of about 3 cm)
- 1 N weight hanger and 1 N weights
- half-metre rule
- string
- 2 clamps, bosses and stand

Plan

- Hang the top ring of the spring on the bar of a clamp in a stand and tie a length of string (about 60 cm) to the bottom ring. Tie a loop on the free end of the string to take the weight hanger.
- Clamp the half-metre rule so that it lies along the links of the spring, with its scale going downwards.
- Read the position of the bottom of the lower ring of the spring on the rule.
- Hook the hanger to the loop on the end of the string and read the new position of the lower ring.
- Be aware of what could happen if the stand fell over or the knots slipped.
- Add the 1 N weights one at a time, reading the position of the lower ring each time.

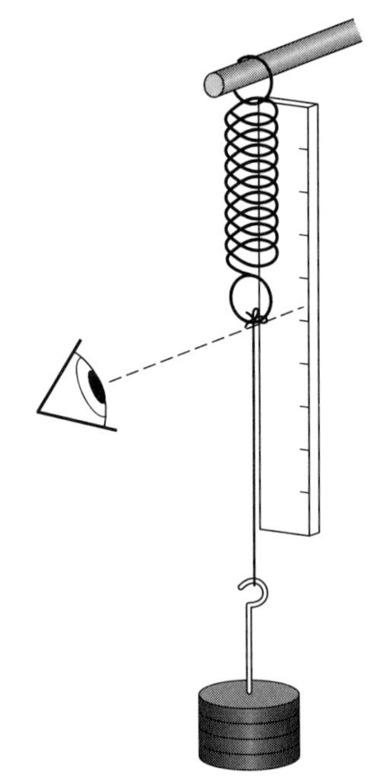

Figure 7.1

Analysis

1 Record your readings in a table and for each load calculate the extension of the spring (scale reading − scale reading with zero load).
2 Plot a graph of force (y-axis) against extension (x-axis) and draw the best-fit line.
3 If you can draw a straight line through the points and the line goes through the origin, conclude that the spring follows Hooke's law over this range of force. Otherwise, conclude that it doesn't. Look out for a slight curve in the line for small extensions as the links of the spring separate.

Sample readings

stretching force /N	0.0	1.0	2.0	3.0	4.0	5.0	6.0	7.0
scale reading /cm	2.0	4.0	7.4	10.9	14.5	16.9	21.0	25.0
extension /cm	0.0	2.0	5.4	8.9	12.5	14.9	19.0	23.0

Plot the graph suggested in the *Analysis* for these sample readings. Decide if there are any anomalous points and whether a little extra force is needed to open the spring in the early stages of the stretch.

Experiment: Measuring the elastic limit and spring constant of a spring

Apparatus As for the Preliminary work (see opposite).

Plan

- Follow the plan for the Preliminary work, then continue to add newton weights until the spring has a small permanent stretch. (You can check this by supporting the weights with your hand.)
- Remove the weights one by one and record the scale reading of the bottom ring as you 'unload'.

Analysis

1. Record your readings in a table that has space for 'loading' and 'unloading' scale readings and extensions (see the *Sample readings* below).
2. Plot graphs of force (y-axis) against extension (x-axis) for the loading and unloading readings, on the same axes. Use a ruler to draw the line of best fit where the points seem to lie in a straight line and draw smooth curves where they do not.
3. Mark the *elastic limit* (where the line begins to leave the straight section).
4. Measure the gradient of the straight section when the spring is being loaded. This is the *spring constant*. Give your result in N m^{-1}.

Sample readings

force /N	0.0	1.0	2.0	3.0	4.0	5.0	6.0	7.0	8.0	9.0	10.0	11.0	12.0	13.0	14.0
scale reading loading /cm	4.2	7.0	11.1	15.2	19.3	23.3	27.3	31.3	35.3	39.3	43.3	47.9	53.0	59.5	66.0
extension loading /cm															
scale reading unloading /cm	13.2	17.7	22.3	26.5	30.9	34.9	38.7	42.5	46.3	49.8	53.2	56.2	59.9	63.0	66.0
extension unloading /cm															

Calculate the extensions for these sample readings and plot the graph suggested in the *Analysis*.
Calculate the spring constant.
Was the spring loaded past its elastic limit?

Answers ~24.5 N m^{-1}; yes.

Investigation: Measuring the spring constant of springs in parallel and in series

Apparatus As for the Preliminary work (see p. 58), plus:

- extra identical springs

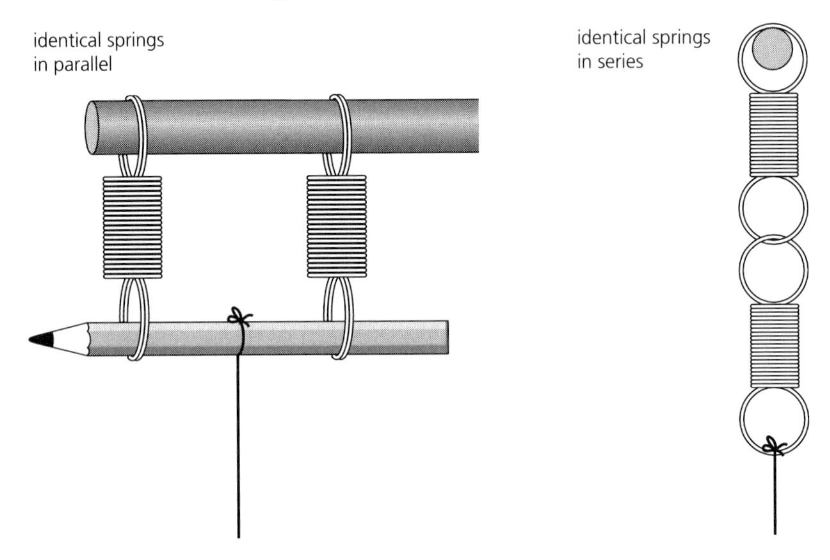

Figure 7.2

Plan

- Use the value of the spring constant found by the previous experiment to predict the spring constant of combined identical springs, (i) in parallel, (ii) in series.
- Use the plan of the previous experiment to measure the spring constant of the springs when they support the load together (in parallel) and when they are joined end to end (in series).

Analysis

1 Design a results table for all your readings.
2 Plot graphs of force against extension for each arrangement. Calculate the spring constant for each arrangement.
3 Report on whether your results support your prediction.

Sample readings

Two equal springs in series

stretching force /N	0.0	1.0	2.0	3.0	4.0	5.0	6.0	7.0
scale reading loading /cm	5.5	9.6	16.9	24.6	32.2	39.2	47.0	54.1
extension /cm	0.0	4.1	11.4	19.1	26.7	33.7	41.5	48.6

Plot a graph to find the spring constant of these two springs in series.

Answer 13.4 N m^{-1}.

Skill level (Analysing)

A: I put results for series and parallel springs into two tables. All columns (or rows) had headings with units. I plotted two load/extension graphs. I calculated the spring constants to the same number of sig. figs as the data. I wrote a conclusion.

All but one of the above = B; all but two = C; all but three = D; all but four = E.

Extension To make this a full investigation, you could take measurements with a third spring in series and in parallel.

7.3 Stretching wires

Preliminary work: Stretching copper wire
Experiment: Measuring the effect of a tensile force on copper wire and finding the Young modulus for copper
Investigation: Measuring the extension of nylon fishing line

Preliminary work

Stretching copper wire

Apparatus

- bare copper wire (SWG 32 or thereabouts)
- 2 pencils
- safety glasses

Plan **Wear safety glasses and make sure other people are well clear of the wire.**

- Cut about 50 cm of the wire. Wind each end securely onto a pencil.
- Pull gently on each end of the wire. Notice that for a small force you can feel (but not see) it 'give' a little, then return to its original length when the force is removed.
- Pull with more force until the wire 'flows' (stretches several centimetres) and stiffens (becomes harder to stretch).
- Finally, exceed the ultimate stress and snap the wire.

Experiment

Measuring the effect of a tensile force on copper wire and finding the Young modulus for copper

Apparatus

- blocks of wood or plastic to clamp the ends of the wire (plastic blocks used to join sheets of MDF, found in DIY stores, work well)
- G-clamp
- pulley wheel on a clamp
- metre rule
- masking tape
- 1 N weight hanger and 1 N weights
- bare copper wire (about SWG 32)
- string
- micrometer screw gauge
- safety glasses
- thick cardboard to protect the floor

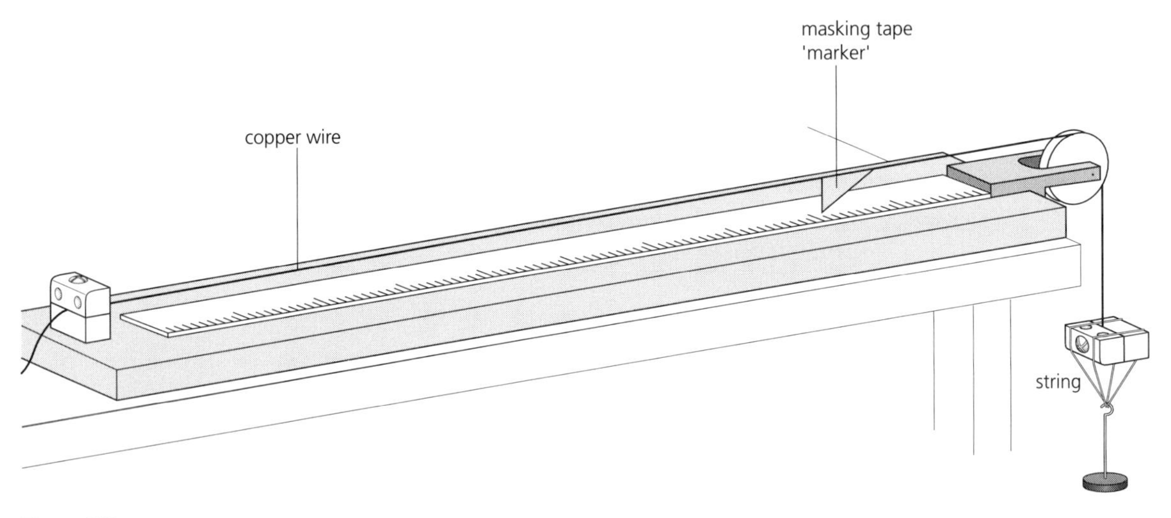

Figure 7.3

Plan **Wear safety glasses and make sure other people stand well away from the wire and the falling weights.**

- Cut a 2–3 m length of copper wire and clamp one end between the plastic blocks. Use the G-clamp to secure the blocks to the end of a long bench.
- Lead the other end of the wire through the pulley so that it hangs over the far end of the bench. Clamp the end of the wire in another pair of blocks and tie a string loop around the blocks to take the weight hanger (see Figure 7.3).
- Stick a small piece of masking tape over the wire about 20 cm from the pulley to act as a marker. Tape the metre rule to the bench so that it lies close to the wire and so that the marker lies over its millimetre divisions (see Figure 7.3).
- Measure the length (L) from the fixed end of the wire to the marker.
- Use the micrometer to measure the diameter of the wire in 5 places, rotating the instrument a little for each reading.
- Hang the weight hanger on the wire to tension it and read the position of the marker on the metre rule for a tensile force of 1 N.
- Gently add 1 N weights to the hanger and read the position of the marker as the wire stretches. When the wire 'flows', wait until it stops before you take a reading.
- Keep increasing the tensile force until the wire breaks.

Analysis

1. Record your readings in a table that includes rows for the *position of the marker* and the *tensile force* (F). Record all your measurements of length L, diameters d and their average.
2. Plot one graph of F against *position of the marker* for the whole range of your readings, and another graph for the elastic (Hooke's law) region only.

3 Measure the gradient of the second graph to obtain an average value of force/extension (F/e).

4 Calculate the average cross-sectional area of the wire (A).

Stress is defined as force per unit area (F/A), *strain* as extension per unit length (e/L), and the *Young modulus* (Y) as stress/strain.

$$Y = \frac{F/A}{e/L}$$

$$Y = \frac{FL}{eA}$$

5 Calculate the Young modulus for copper.

Sample readings

Length of wire = 257 cm
Diameter of wire (SWG 32) /mm = 0.28, 0.28, 0.28, 0.28, 0.28

force /N	1	2	3	4	5	6	7	8	9	10	11	12
scale reading /cm	37.5	38.2	39.0	39.5	40.5	42.0	43.6	45.6	48.2	53.6	80.0	140.0

Plot the two graphs suggested in the *Analysis* for the sample data above.
Find the gradient of the Hooke's law section.
Calculate the cross-sectional area of the wire.
Calculate a value for the Young modulus.

Answer ~5.7×10^{10} Pa.

Skill level (Analysing)

A: I put my force values, scale readings and extension readings into a table. I plotted two graphs with different scales. I identified the elastic region and calculated the gradient of the line. I calculated the cross-sectional area of the wire. I calculated a value for the Young modulus to 2 sig. figs.

All but one of the above = B; all but two = C; all but three = D; all but four = E.

Evaluation

The uncertainty of the micrometer reading will be about 0.01 mm in a reading of 0.20 mm. This is a 5% error that becomes 10% when diameter is squared to obtain cross-sectional area.

In comparison, the uncertainties in F and L will be small (about 2 mm in 2000 mm or 0.1% for L).

A large uncertainty comes from measuring the position of the marker on the metre rule when the stretch is very small. The uncertainty can be estimated most easily from the range of gradients of what could be considered best-fit lines through the plotted points. It is likely to be about 10%.

The overall uncertainty in the value of the Young modulus would therefore be about 20%.

Improving the plan

- Use a vernier and scale to measure the extension. Fix the vernier to the end of the wire so that the vernier slides over a scale that has been taped to the bench. Use a second length of wire to attach the vernier to the weight hanger (see Figure 7.4). This will allow measurements of extension to be made to 0.1 mm.
- Use a conversion table for Standard Wire Gauge to metric (see p. 71). This gives the diameter of the wire in mm to 3 decimal places.

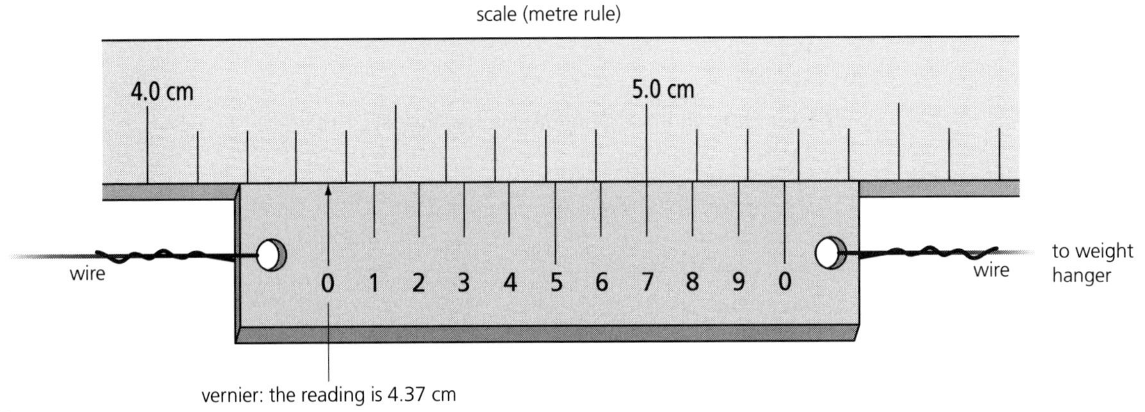

Figure 7.4

Investigation: Measuring the extension of nylon fishing line

The same plan as in the previous experiment can be used to investigate the Young modulus for nylon. Watch the behaviour of the material closely as you apply force, and wait for it to stop 'creeping' (stretching without further increasing the force) before you take a reading.

Sample readings

Diameter = 0.20 mm
Breaking force = 20 N

force /N	0	1	2	3	4	5	6	7	8
scale reading /cm	463.2	466.3	471.2	478.3	484.3	489.2	492.7	495.9	498.5
extension /cm	0.0	3.1	8.0	15.1	21.1	26.0	29.5	32.7	35.3

force /N	9	10	11	12	13	14	15	16
scale reading /cm	502.3	504.0	505.8	507.1	509.5	512.1	515.6	518.0
extension /cm	39.1	40.8	42.6	43.9	46.3	48.9	52.4	54.8

Plot a force/extension graph from this sample data. Draw a smooth curve through the points and notice that nylon does not follow Hooke's law and therefore its Young modulus varies with force.

Draw tangents at force values of 3 N and 8 N and calculate the Young modulus at these values.

Answers ~2470 MPa; ~5500 MPa.

Extension To make this a full investigation, find how the Young modulus of nylon varies with force.

7.4 Stretching polymers

Experiment 1: Plotting a force/extension graph for rubber
Experiment 2: Plotting a force/extension graph for polythene

Experiment 1 Plotting a force/extension graph for rubber

Apparatus

- rubber bands
- 1 N weight hanger and extra 1 N weights
- clamp stand and attachments
- metre rule
- strong thread

Plan

- Hang the rubber band from the arm of a clamp in a stand.
- Tie a 60 cm length of strong thread to the bottom of the band and make a loop at the free end of the thread for the weight hanger.
- Fix the metre rule in a second clamp so that it lies along the length of the rubber band.
- Flatten the band against the rule and record the position of the bottom of the rubber band on the rule when there is no tensile force.
- Hang the weight hanger on the thread and record the new position of the bottom of the rubber band.
- Add the 1 N weights one at a time and, for each one added, record the position of the bottom of the rubber band. You should wait until the rubber band has stopped 'creeping' (stretching a long way without increasing the tensile force) before you take a reading.
- Take the 1 N weights off, one at a time, and record the position of the bottom of the band as the hanger is unloaded.

Analysis

1. Put your results into a spreadsheet or a table, with rows for force (loading and unloading), scale reading and extension.
2. Plot a graph of force (y-axis) against extension (x-axis) for loading and unloading, and join the points with smooth lines. Label the steeper sections of the line 'hard to stretch' and the less steep sections 'easy to stretch'. If you judge there to be a Hooke's law section, where the line is straight and goes through the origin, mark this too.

Sample readings

	loading							unloading						
force /N	0.0	1.0	2.0	3.0	4.0	5.0	6.0	7.0	6.0	5.0	4.0	3.0	2.0	1.0	0.0
scale reading /cm	1.2	2.0	4.1	7.6	12.9	17.5	21.4	24.7	22.9	20.3	16.0	10.8	6.4	3.1	1.2
extension /cm	0.0	0.8	2.9	6.4	11.7	16.3	20.2	23.5	21.7	19.1	14.8	9.6	5.2	1.9	0.0

Use the sample readings to plot the graph described in the *Analysis*.

Answer There is no Hooke's law region.

Experiment 2 Plotting a force/extension graph for polythene

Apparatus

- strips of polythene 1 cm × 20 cm cut from a plastic bag with a scalpel
- 2 pairs of plastic connector blocks joined by a nut and bolt
- metre rule
- 1 N weight hanger and 1 N weights
- strong thread
- clamp, boss and stand
- sheets of thick cardboard to protect the floor

Plan

- Grip each end of the polythene strip in a pair of plastic blocks.
- Fix the top pair of blocks in a clamp on a stand and tie a length of strong thread to the bottom pair.
- Clamp the metre rule so that the bottom pair of blocks lie against its scale (see Figure 7.5).
- Note the position of the bottom blocks on the scale, then hang the weight hanger onto the end of the thread.
- Read the position of the bottom blocks for this force of 1 N.
- Add 1 N weights and measure the position of the bottom blocks each time. At some stage the plastic will 'creep'. You should wait for this to stop before taking a reading.
- Continue to increase the tensile force until the strip breaks.

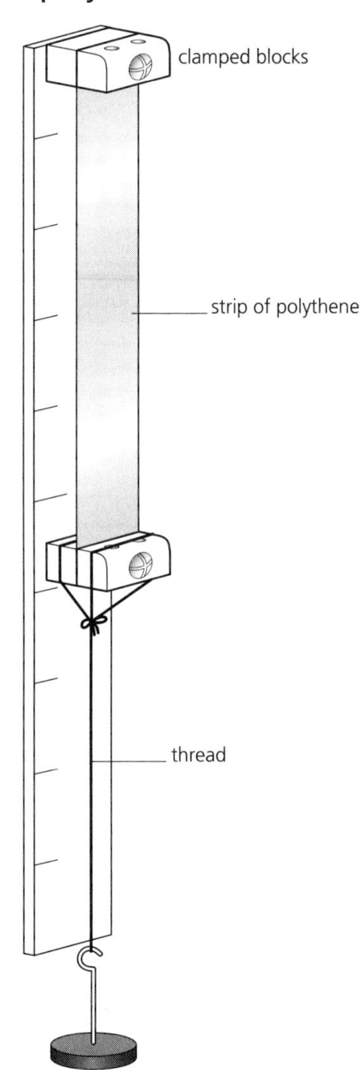

Figure 7.5

Analysis

1. Put your results into a table, leaving a row for calculating the extension.
2. Plot a graph that shows the behaviour of polythene when it is stretched. If you use a spreadsheet you can enter a formula to do the calculations and plot the graph automatically.

Sample readings

force /N	0.0	1.0	2.0	3.0	4.0*	5.0**
scale reading /cm	14.8	15.0	15.2	15.6	32.1	60.0
extension /cm	0.0	0.2	0.4	0.8	17.3	45.2

*A neck formed and then about half the length thinned down gradually, causing a large extension.
**The rest of the strip thinned down, stretching as it did so. The strip then snapped.

Plot the graph suggested in the *Analysis* for these sample readings. Compare the behaviours of rubber and polythene when they are stretched by a tensile force.

8 Direct current electricity

8.1 Electrical measurements

Preliminary work 1: Measuring resistance using an ohmmeter
Experiment 1: Measuring resistance using an ammeter and a voltmeter
Experiment 2: Measuring the resistivity of Nichrome
Preliminary work 2: Using an oscilloscope to measure pd
Experiment 3: Measuring the peak-to-peak pd of a source of alternating voltage

Preliminary work 1 Measuring resistance using an ohmmeter

Apparatus

- Nichrome resistance wire (SWG 32 or equivalent)
- wire cutters
- selection of multimeters with resistance ranges
- connecting leads with crocodile clips
- metre rule
- adhesive tape

Plan

- Cut just over a metre of the Nichrome wire and tape it along the length of a metre rule.
- Switch the multimeter to its lowest resistance range (e.g. 200 Ω) and check for a zero error by connecting a copper lead across its terminals. If the reading is not zero, there is a zero error that should be subtracted from all readings.
- Clip leads to each end of the wire so that the crocodile clips are exactly 1 m apart. Connect the leads to the multimeter and record the resistance value.
- Without moving the clips, obtain readings of the zero error and the wire's resistance using other meters.

Analysis

1 Record the readings from each meter and work out an average.
2 Write a conclusion from your findings.

Sample readings

zero error /Ω	resistance reading /Ω	corrected for zero error /Ω
0.0	17.9	17.9
0.1	18.0	17.9
0.1	17.9	17.8
0.0	18.0	18.0
0.0	17.9	17.9

Average resistance = 17.9 ± 0.05 Ω

Evaluation

Uncertainties in the value of the resistance can arise from:

- the calibration of the meters and their resolution;
- contact resistance at either end of the copper leads;
- change in temperature (although this will be very small).

The variation of the sample readings shows that the meters agree very closely, with a maximum difference of about 0.1 Ω between their readings.

Experiment 1 Measuring resistance using an ammeter and a voltmeter

Apparatus

- 1 m of Nichrome resistance wire (from Preliminary work 1)
- low voltage dc power supply
- ammeter (or multimeter)
- voltmeter (or multimeter)
- connecting leads with crocodile clips

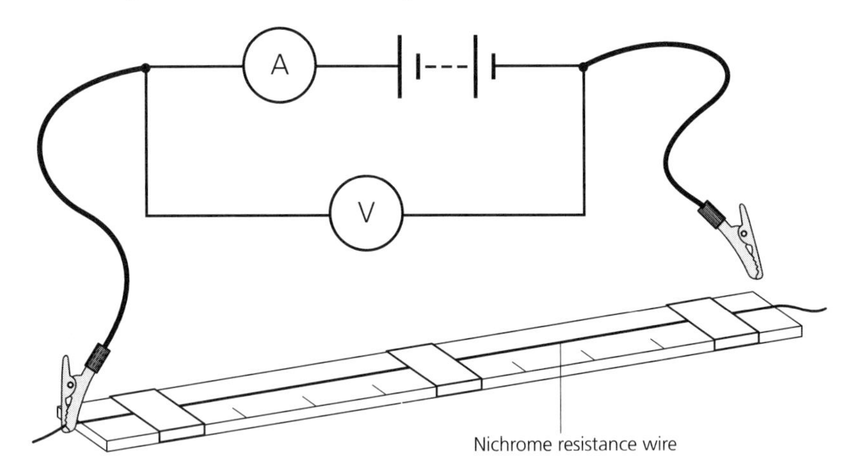

Figure 8.1

Plan

- Join the power supply, the ammeter and one end of the resistance wire in series. Leave the other end of the wire unconnected.
- Add the voltmeter, switch on the power supply and, before connecting the crocodile clip, adjust the emf to about 3 V.
- Complete the circuit briefly by touching the crocodile clip on the wire at the 100 cm mark.
- Record the ammeter and voltmeter readings.

Analysis

1. Write down the readings to the number of decimal places shown by the meters.
2. Calculate resistance (= pd/current).
3. Compare with the results from Preliminary work 1, and comment on the two ways of measuring resistance.
4. Would it have made any difference if the voltmeter had been connected to the crocodile clips instead of to the other ends of the connecting leads?

Sample readings

pd /V	2.15
current /mA	120

Calculate the resistance from these sample readings.

Answer 17.9 Ω.

Experiment 2 Measuring the resistivity of Nichrome

Apparatus As for Experiment 1, plus:

- micrometer

Plan

- Build the circuit and measure the current through and pd across 100 cm of the wire, as in Experiment 1.
- Record the length of wire (L) between the crocodile clips and leave the wire a short time to cool down.
- Touch one clip on the wire at the 90 cm mark and record current, pd and length again.
- Continue with shorter and shorter lengths, allowing the wire to cool between readings. Obtain 6 values of current, pd and length in this way.
- Check the micrometer for a zero error then use it to measure the diameter of the wire. Do this at 5 places along its length (in case it tapers), rotating the micrometer as you go (in case the wire is not circular).

Skill level (Implementing)

A: I built the circuit correctly and without help. I took readings every 10 cm going down from 100 cm. I connected the crocodile clip briefly while I read the meters. I left the wire to cool before I moved on. I did not burn myself on the wire or damage the meters.

All but one of the above = B; all but two = C; all but three = D; all but four = E.

Analysis

1 Record your readings in a spreadsheet or a clearly labelled table.

2 Calculate the average diameter (d) and cross-sectional area (A) of the wire ($A = \pi d^2/4$).

3 Calculate the resistance (R) for each length of wire from the pd and current values.

Resistivity is $\rho = \dfrac{RA}{L}$

so $R = (\rho/A)L$

A graph of R (y-axis) against L (x-axis) should therefore be a straight line through the origin with a gradient of (ρ/A).

4 Plot this graph, measure its gradient and multiply by A to obtain ρ.

Sample readings

Nichrome wire SWG 32

Diameter = 0.274 mm (taken from conversion tables, see p. 71)

length /m	1.0	0.9	0.8	0.7	0.6	0.5	0.4	0.3	0.2	0.1
pd /V	5.00	4.97	4.91	4.85	4.78	4.71	4.55	4.38	4.11	3.49
current /mA	278	303	342	380	442	499	621	780	1080	1700
resistance /Ω	18.0	16.4	14.4	12.8	10.8	9.4	7.3	5.6	3.8	2.1

Plot the graph of resistance against length for these readings, and use it to obtain a value for resistivity.

Answer ~$1.06 \times 10^{-6}\,\Omega$ m.

Evaluation A large source of uncertainty is in measuring the diameter of the wire. The micrometer reads 0.01 mm but the diameter is only around 0.3 mm, giving an uncertainty of about 3%. This is doubled to 6% when the diameter is squared to calculate A.

The uncertainty in R/L can be obtained from the gradients of the possible lines that can be drawn through the points of the graph (~4% with the sample results), giving an overall uncertainty of around 10%.

Improving the plan

- Use this 'Standard Wire Gauge (SWG) to metric' conversion table to look up the diameter of the wire. Use the internet to find values outside the range shown.

SWG	21	22	23	24	25	26	27	28	29	30	31	32
diam. /mm	0.813	0.711	0.610	0.559	0.508	0.457	0.417	0.378	0.345	0.315	0.295	0.274

Preliminary work 2 Using an oscilloscope to measure pd

Apparatus

- oscilloscope
- selection of cells and batteries (e.g. PP3, AAA)
- dc/ac variable voltage supply
- leads to fit the input sockets of the oscilloscope

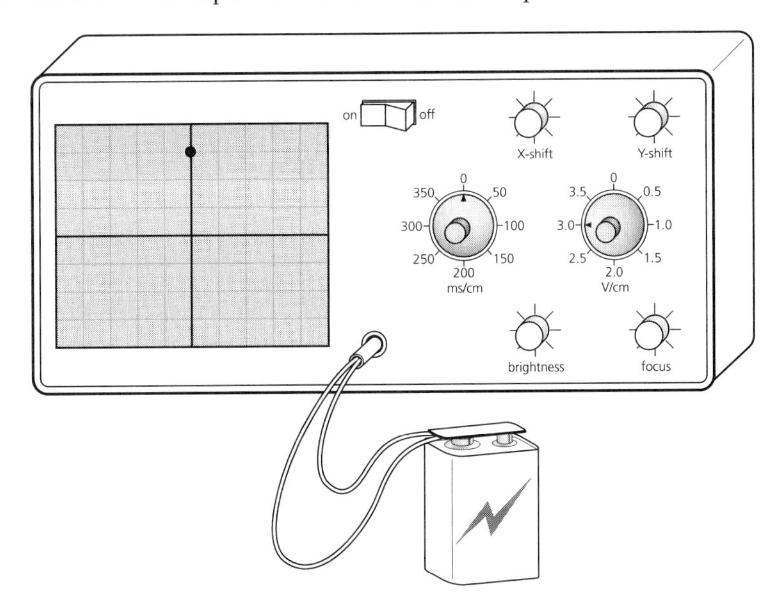

Figure 8.2

Plan

- Switch on the oscilloscope, turn off its *time base* (ms/cm or ms/div) and use the *X-shift* and *Y-shift* controls to position the spot at the centre of the screen. Use the *brightness* and *focus* controls to make the spot as small and sharp as possible. Do not have the spot too bright when it stays in one position on the screen.

- Connect the terminals of one of the batteries to the *input* socket of the oscilloscope. The spot should jump up or down.
- Adjust the *Y-amplifier sensitivity* (V/cm or V/div) to make the jump as large as possible. Use the centimetre grid to measure the deflection. (Make sure the Y-amplifier *fine control* is set to 'calibration' or turned fully to the left. The sensitivity settings are then as they are shown on the markings.) Record the deflection (in cm) and the Y-amplifier sensitivity setting (in V/cm).
- Repeat for the other batteries.

Analysis Make a table with columns for:

- the source of the emf;
- the deflection (cm);
- the Y-amplifier sensitivity setting (V/cm);
- the calculated pd (deflection × sensitivity setting).

Evaluation The graph grid of an oscilloscope is drawn on a clear plastic screen that lies in front of the vacuum tube. The spot is formed on the inside of the vacuum tube and there could be parallax between the spot and the grid that will cause an error in reading its position. The scale of the grid is usually marked in 2 mm divisions and the spot will have a finite size. This makes it difficult to read its position accurately. The uncertainty in the displacement could be up to 2 mm out of about 5 cm (= 4%).

Oscilloscopes are carefully designed to give a linear response but they cannot be read with great precision. Their greatest use is in viewing the shape of changing voltage signals.

Experiment 3 Measuring the peak-to-peak pd of a source of alternating voltage

Apparatus As for Preliminary work 2, plus:

- power pack with an ac output
- multimeter or ac meter

Plan

- Plug the ac output from the voltage supply into the oscilloscope. This will give a vertical line on the screen (with the time base off).
- Adjust the Y-amplifier control until the line on the grid is as long as possible.
- Record the length of the line (in cm) and the Y-amplifier sensitivity setting (in V/cm).

Analysis

1 Calculate:

- the peak-to-peak pd (Y-amp sensitivity setting × line length);
- the peak pd (half the peak-to-peak pd);
- the root mean square (rms) value (peak pd/$\sqrt{2}$).

2 Compare this rms value of the pd with the value measured by an ac meter or multimeter.

8.2 Voltage/current characteristics

Experiment: Measuring the current that flows in a resistance wire for different pds
Investigation 1: Determining the voltage/current characteristic of the filament of a low voltage lamp
Investigation 2: Determining the voltage/current characteristic of a semiconductor diode

Experiment

Measuring the current that flows in a resistance wire for different pds

Apparatus

- bare Nichrome resistance wire (SWG 32 or equivalent)
- (digital) ammeter
- (digital) voltmeter
- continuously variable dc power supply
- terminal block connectors
- connecting leads
- switch
- heat-resistant mat

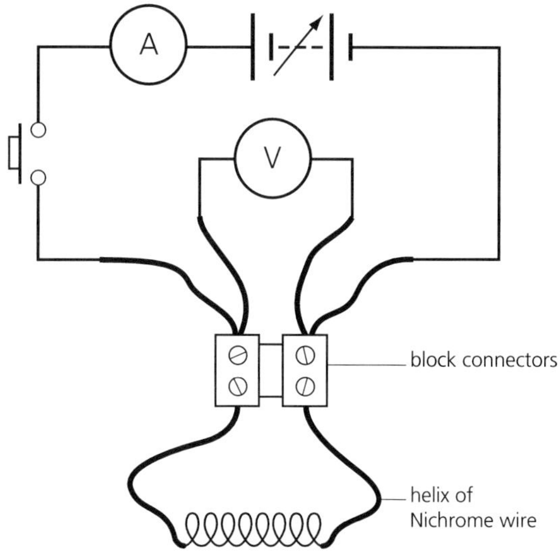

Plan

- Wind a metre of Nichrome wire onto a pencil to form a helix. Use a connecting block to attach two connecting leads to each end of the resistance wire – one for the voltmeter and the other for the main circuit (see Figure 8.3).
- Build the circuit with the power supply, ammeter, resistance coil (on a heat-resistant mat) and switch in series. (You could use the switch on the power supply but it may take the voltage some time to drop to zero.) Add the voltmeter across the resistance wire to measure the pd.
- Do a rough check to establish the range of voltage readings you are going to use and to ensure that the ammeter gives suitable readings. From the check, decide on how many readings to take and the interval between them.
- Leave the wire to cool to room temperature, then switch on briefly and take pd and current readings. Change the supply voltage and continue to take readings over the range you have chosen. Leave the wire to cool between readings. It is better to take a large number of separate readings than to take repeat readings because of the difficulty of re-setting the voltage to a previous value.

Figure 8.3

Skill level (Implementing)

A: I built the circuit without any mistakes or short circuits. I did a rough check to establish ranges. I got 6 pairs of readings of pd and current in increasing size and tabulated them clearly. I switched off after each reading to let the wire cool. I did not overload the meters or burn myself on the wire.

All but one of the above = B; all but two = C; all but three = D; all but four = E.

Analysis

1 Put your readings into a headed table. Calculate a resistance value from each pair of pd and current values.
2 Plot a graph of current against pd. Draw the best-fit line and, if you decide it is straight, measure the gradient to get an average value of resistance.
3 Write a conclusion saying whether your work convinces you that Nichrome resistance wire follows Ohm's law. If the graph is a straight line through the origin, then it shows that *pd and current are directly proportional provided temperature is constant*; the individually calculated resistance values will show you whether or not the resistance is constant.

Skill level (Analysing)

A: I plotted a large-scale, high quality graph, with clear labels and units. The points were plotted accurately and the best-fit line well chosen. I measured the gradient using a large triangle. I wrote a clear conclusion which included a statement of Ohm's law. All results were given to the correct number of significant figures.

All but one of the above = B; all but two = C; all but three = D; all but four = E.

Sample readings 1 metre of Nichrome wire SWG 32

pd /V	0.30	0.84	1.39	1.74	2.39	2.95	3.75	4.27	5.36	6.14	7.05	8.23	9.16	10.14
current /mA	16	47	78	97	133	165	209	238	299	342	392	458	510	564
resistance /Ω	18.8	17.9	17.8	17.9	18.0	17.9	17.9	17.9	17.9	18.0	18.0	18.0	18.0	18.0

average	18.0
std	0.2

Plot the graph suggested in the *Analysis* above for these sample readings, and calculate the resistance of the wire from the gradient.

Answer ~18.0 Ω.

Evaluation The uncertainties in pd and current depend on the accuracy of the calibration of the meters and on whether the temperature changes. This can happen easily if you let the current run through the wire for too long.

The uncertainty in the resistance can be estimated as half the biggest difference of the readings from the average, or from the standard deviation (0.2 Ω or 1% for the sample readings).

Investigation 1 Determining the voltage/current characteristic of the filament of a low voltage lamp

Apparatus Circuit as in the previous experiment, plus:

- 12 V 24 W tungsten filament lamp (e.g. ray box lamp)

Plan

- Switch off the power supply and replace the resistance wire with the lamp.
- Do a rough check by increasing the voltage from 0 to 12 V to make sure that the ammeter is suitable for the job. Take care – the lamp will get hot.
- Let the lamp cool down and then obtain readings of pd and current as the supply voltage is increased from zero.
- Make observations of the colour and brightness of the filament as you go, but take care not to stare directly into the bright lamp.

Analysis

1 Put your readings into a table with columns or rows for:
pd /V; current /A; pd ÷ current /Ω; filament colour and brightness.

2 Calculate the resistance of the filament for each pd value.
3 Plot a large-scale graph of pd (y-axis) and current (x-axis). Draw a best-fit line or curve through the points, ignoring any that you consider to be anomalous.
4 Write your conclusion. A simple conclusion would be that either pd and current are proportional or they are not, and that resistance is or is not constant. A more discerning conclusion would be to state qualitatively how resistance changes with pd.

Alternative analysis A higher level analysis would be to recognise that the resistance increases with the rise in temperature caused by the heating effect of the current. The change in resistance will then alter the current that caused the change. The equation relating current and pd is therefore likely to be complex.

Theory shows that if the filament behaves like a perfect (black body) radiator, current (I) and pd (V) are connected by the equation:

$$I = aV^b$$

where a and b are numerical constants.

If you take logs (to base e), this equation becomes:

$$\ln I = \ln a + b \ln V$$

which compares with the equation of a straight line:

$$y = c + mx$$

To test this equation for the filament, you need to plot a graph of $\ln I$ against $\ln V$. A straight line will confirm your assumption and give you b (the gradient) and $\ln a$ (the intercept on the y-axis).

Your final conclusion could be a detailed equation linking I and V and a statement of the range over which the equation applies.

Sample readings

pd /V	0.00	0.45	1.10	1.87	2.68	3.57	4.46	5.32	6.32	7.29	8.25	9.24
current /A	0.00	0.33	0.62	0.78	0.92	1.06	1.20	1.31	1.44	1.55	1.65	1.76
resistance /Ω		1.36	1.78	2.40	2.91	3.37	3.71	4.06	4.39	4.70	5.00	5.25

Plot the log graph suggested in the *Alternative analysis* (p. 75) for these sample readings, and find values for the constants a and b.

Answers $b \approx 2.02$; $a \approx 0.58$.

Evaluation The log graph will tell you immediately whether or not the theoretical equation describes the behaviour of the filament lamp. The scatter of points above and below the line indicates the size of the uncertainties in each pair of values. The uncertainty in a and b can be estimated from the extreme lines that could go through the points.

Improving the plan

- Take readings at smaller voltage intervals when the current is small, so that you can plot the graph more accurately.

Extension To make this a full investigation, find out for yourself if the current and pd for a filament lamp are related by a power law as suggested in the *Alternative analysis*.

Investigation 2 Determining the voltage/current characteristic of a semiconductor diode

Apparatus Circuit as used previously, plus:

- silicon diode or light emitting diode (LED)
- resistor (330 Ω)

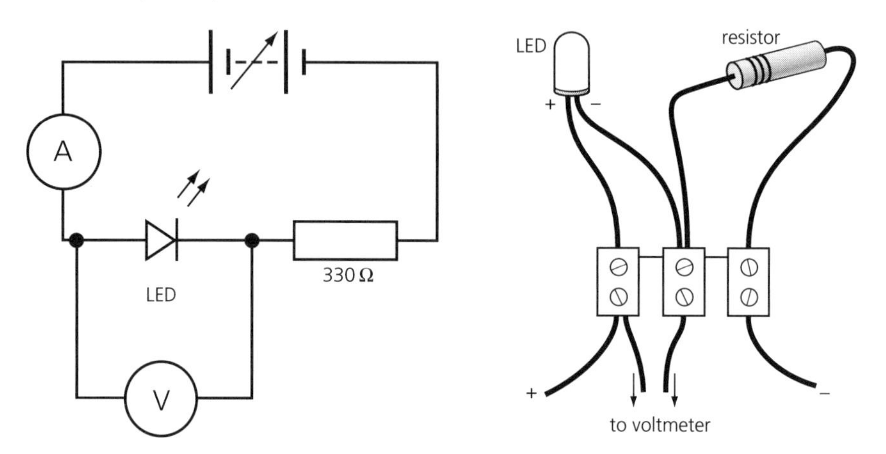

Figure 8.4

Plan

- Connect the diode and its protective resistor into the circuit, and the voltmeter across the diode. Make sure the diode is forward biased (with the silicon diode, the end with the band should be further from the + of the power supply).

- Do a rough check to see how the current changes with voltage. Start the pd from zero and increase it very gently, especially in the region where the current starts to flow. Adjust the current range on the ammeter to give the most precise readings for the pd values you are going to use. Decide on the range and number of readings to take.
- Go back and record detailed measurements of pd and current up to about 3 V. Then take some extra readings at interesting points where there is a rapid current change.
- Reverse the diode and check that no measurable current flows within this range of voltage.

Analysis

1. Put your readings into a table with an extra row for calculating resistance.
2. Plot a graph of pd against current. Take special care over the scale of the x-axis.
3. Identify a straight section of the graph and use the pd and current values in the table to calculate the resistance over that section.
4. Write a few sentences describing the *V/I* characteristic and stating whether a diode follows Ohm's law.

Sample readings

LED + 330 Ω

pd /V	0.21	0.37	0.62	0.73	0.89	1.00	1.21	1.38	1.58
current /mA	0.00	0.00	0.00	0.00	0.00	0.00	0.00	0.03	6.40

pd /V	1.59	1.62	1.63	1.65	1.66	1.67	1.68	1.70	1.77
current /mA	8.70	15.30	25.90	35.90	43.40	50.30	66.20	80.00	88.50

Extra readings taken around a sudden change in current

pd /V	1.47	1.45	1.50	1.55	1.57
current /mA	0.39	0.31	1.17	3.78	5.66

Plot the graph suggested in the *Analysis* for these sample readings.

Evaluation

A diode has a very high resistance for pds up to about 0.6 V (for a silicon diode) or 1.2 V for the LED used for the sample readings. Then it suddenly starts to conduct. Its resistance then becomes very low (~1.5 Ω for the sample readings). The series resistance is necessary to protect the diode, power supply and ammeter from a near short circuit.

Improving the plan

- Take even more closely spaced readings around the 'turn-on' point to give more detail of the rapid changes that take place there.

Extension

To make this a full investigation, you could compare the voltage/current characteristics of different types of diode.

8.3 The resistance of a thermistor

Experiment: Measuring the change in resistance of a thermistor with temperature
Full investigation: Does a thermistor follow Ohm's law at different temperatures?

Experiment — Measuring the change in resistance of a thermistor with temperature

Apparatus

- thermistor (e.g. one with resistance ~1 kΩ at 0 °C and a negative temperature coefficient)
- multimeter with resistance range
- beaker of water
- Bunsen, tripod and gauze
- clamp and stand
- electronic or mercury thermometer (0–110 °C)
- stirrer
- crushed ice
- safety glasses

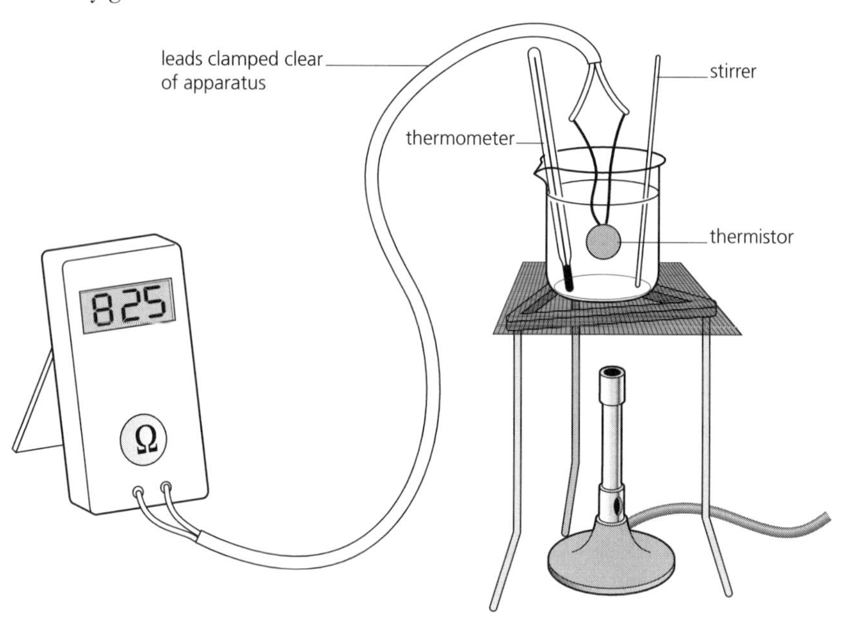

Figure 8.5

Plan

- Put the beaker of water on the tripod and gauze, and add enough ice to cool the water to below 5 °C. Stir until the ice has melted.
- Connect the thermistor to the multimeter and select the range that gives the most precise reading of resistance. Check the multimeter for zero error (see p. 68).
- Submerge the thermometer and the thermistor in the cold water. Use a clamp and stand to hold the leads away from the Bunsen flame that you will use shortly (and away from the gauze which will get very hot).

- Wait until the thermistor is at the temperature of the water (this will be when the resistance value stabilises) and record the resistance and the temperature.
- Raise the temperature of the water with the Bunsen by about 10 degrees. Remove the Bunsen, stir thoroughly until the thermometer reading is steady, then read the temperature and resistance again.
- Continue every 10 degrees until the water boils.

Analysis

1. Record your readings in order of increasing temperature in a headed table.
2. Plot a large-scale graph of resistance against temperature, drawing a smooth, best-fit curve through the points. Look for anomalous points and ignore them when you draw the line.
3. Write a conclusion describing in detail how the resistance changes, including whether the rate of change alters.

Alternative analysis

A higher level analysis would be to see if resistance (R) and temperature (θ) are related by an equation of the form $R = a\theta^b$, where a and b are numerical constants.

Taking logs to base e, this equation becomes:

$$\ln R = b \ln \theta + \ln a$$

which compares with the equation of a straight line:

$$y = mx + c$$

If the equation applies, a graph of $\ln R$ against $\ln \theta$ will be a straight line. The gradient will be the power, b, and the intercept will be $\ln a$.

Plot the graph and decide if the equation applies. If it does, use the graph to obtain values for b and a. Conclude by quoting an empirical equation that can be used to calculate the resistance of the thermistor at any temperature.

Sample readings

resistance /Ω	822	562	438	380	194	138	102	82	65	59	52	49
temperature /°C	4.8	14.6	21.5	29.3	37.2	48.0	59.2	67.5	79.3	87.0	91.6	100.9

Plot a graph of resistance against temperature for the sample data.

Answer There is an anomalous point at 29.3 °C.

Plot the graph suggested in the *Alternative analysis* above.

Answer The equation $R = a\theta^b$ applies for temperatures above 21 °C.

Evaluation

The largest uncertainty is in the temperature value because the thermistor is not likely to be exactly at the temperature of the water. The water, unless thoroughly mixed, will vary in temperature from point to point and the thermometer cannot be in the same place as the thermistor. Reading the thermometer from a distance through glass and water will introduce parallax and reading uncertainties of at least one degree. There will also be calibration uncertainties.

You can expect at least a 2% uncertainty in the resistance value given by the meter.

The size of these uncertainties can be judged from the scatter of the points above and below the best-fit line. It should be possible to judge which points are due to anomalies rather than random uncertainties.

Improving the plan

- Start with the thermistor in water at boiling point and take the readings as the water cools. This will give the thermistor longer to adjust to the water temperature and there will be less turbulence. There is also less risk of an accident involving the Bunsen flame during the experiment. Alternatively, you could add boiling water from a kettle in stages to heat the water.
- Use an electronic thermometer with a digital readout and attach its probe to the body of the thermistor.

Full investigation

Does a thermistor follow Ohm's law at different temperatures?

Apparatus

- (continuously) variable dc power supply
- ammeter (or multimeter)
- voltmeter (or multimeter)
- switch
- thermistor
- crushed ice in a filter funnel standing in a beaker
- 2 beakers of water
- Bunsen, tripod and gauze
- electronic or mercury thermometer
- safety glasses

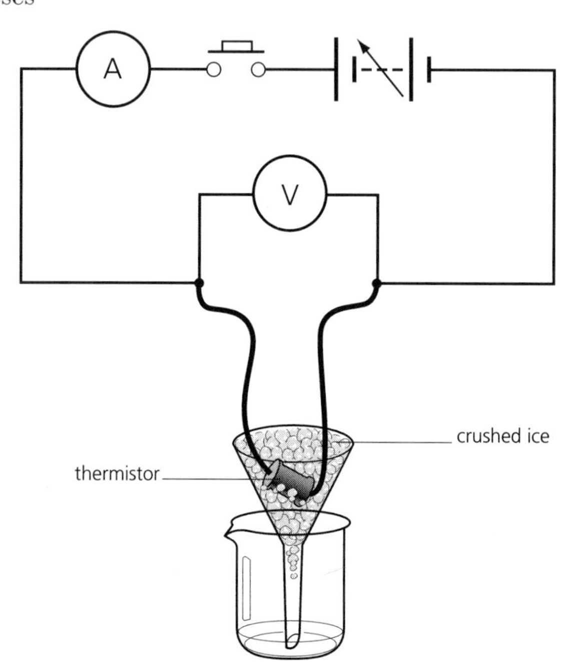

Figure 8.6

Plan The idea is to measure pd and current values for the thermistor at three different temperatures.

- Build a circuit with the thermistor, dc supply, switch and ammeter in series.
- Add the voltmeter across the thermistor to measure its pd.
- Do a rough check by increasing the supply voltage to about 10 V and adjusting the ammeter range to give precise current readings.
- Submerge the thermistor in the ice in the funnel. Wait until the ice is melting and water is dripping out of its spout. The thermistor will then be at a constant temperature.
- Start with a small voltage and switch on briefly while you take the pd and current values.
- Increase the supply voltage in small steps and repeat until you reach your chosen maximum.
- For the next temperature, put the thermistor into pure boiling water. Keep the water boiling while you get a set of pd and current readings for this temperature.
- For a third temperature, put the thermistor into a beaker of water that has been standing in the room for some time. Get a set of pd and current readings as before and record the temperature.

Skill level (Implementing)

A: I built the circuit correctly without help. I changed the voltage and measured 6 values of pd and current with the thermistor at 0 °C. I switched on briefly to take the readings and allowed the thermistor to cool between readings. I heated the thermistor in boiling water safely. I got readings for the other two temperatures.

All but one of the above = B; all but two = C; all but three = D; all but four = E.

Analysis

1 Put all your readings into a table on one side of a sheet of paper. For each temperature you will need columns for:

current /mA; pd /V; pd ÷ current /Ω

2 Plot pd against current for the three sets of readings on one graph (if this doesn't compromise the scales).
3 Draw best-fit straight lines if your points justify them; the points should be equally scattered on both sides of the line.
4 Write your conclusion. A simple conclusion would be that the thermistor does follow Ohm's law at all three temperatures. A more perceptive conclusion would be that, within the uncertainties of the measurements, the thermistor follows Ohm's law; however, at the higher temperatures where the uncertainties are likely to be greater (see *Evaluation* overleaf), there could be a departure from proportionality.

Sample readings

Temperature = 0 °C

pd /V	current /mA	resistance /Ω
0.52	0.60	867
1.31	1.61	814
2.19	2.69	814
2.91	3.60	808
3.95	4.91	804
4.59	5.52	832
5.01	6.22	805
5.74	7.00	820

Average resistance = 821 Ω

Temperature = 27 °C

pd /V	current /mA	resistance /Ω
0.97	3.91	248
2.37	9.92	239
3.27	14.89	220
4.07	19.61	208
5.13	25.90	198
5.90	28.92	204
6.62	32.11	206
7.20	36.61	197

Average resistance = 215 Ω

Temperature = 100 °C

pd /V	current /mA	resistance /Ω
0.51	16.3	31.3
1.23	45.1	27.3
1.79	60.0	29.8
2.53	87.5	28.9
3.39	114.0	29.7
4.45	135.6	32.8
5.44	165.9	32.8
6.62	209.0	31.7

Average resistance = 30.5 Ω

Use these sample readings to plot pd against current for all three temperatures on the same graph.

Decide if the thermistor does appear to follow Ohm's law. Explain how this can be confusing when thermistors are designed to have a resistance that changes with temperature.

Evaluation

Despite your best efforts, it is unlikely that the temperature will remain fixed to within one degree. The thermistor resistance alters rapidly with temperature, especially at lower temperatures. You can get a guide to the overall uncertainty from the variation of the resistance values from their average. For example, using the sample results, at 0 °C, the variation of resistance from the average is about 14 Ω. This is 14/821 = 1.71%, which rounds up to 2%. Any non-linear behaviour will lie within this limit and you can quote an uncertainty of about 2% for this temperature. Similar calculations show uncertainties of 7% at 27 °C and 5% at 100 °C.

The voltmeter may introduce a systematic error, because it is connected in parallel with the thermistor and its circuitry may take a small current from the supply. The effect would be to make the measured resistance of the thermistor less than it really is.

Improving the plan

- Temperature control is the key to doing an accurate test, so leave the thermistor for a long time at the chosen temperature before you start taking readings. At 0 °C, make sure the thermistor is completely covered with finely crushed ice leaving plenty on top to allow for melting. Leave them together for long enough for water from the melting ice to be dripping freely from the funnel.
- Make sure you only switch on the current for a brief time while you read the meters. The current will generate heat and alter the temperature of the thermistor by a small amount.
- Uses a voltmeter with a very high resistance.

8.4 The internal resistance of a source of emf

Preliminary work: Measuring the voltage 'drop' when a cell drives a current
Computer simulation: InternalR
Experiment: Measuring the internal resistance of a 'dry' cell

Preliminary work — Measuring the voltage 'drop' when a cell drives a current

Apparatus

- circuit board
- 'dry' cell (e.g. zinc chloride C or D type)
- switch
- 3 lamps (e.g. 6 V 0.3 A) in holders
- voltmeter (or multimeter on dc volts)
- elastic band or cell holder

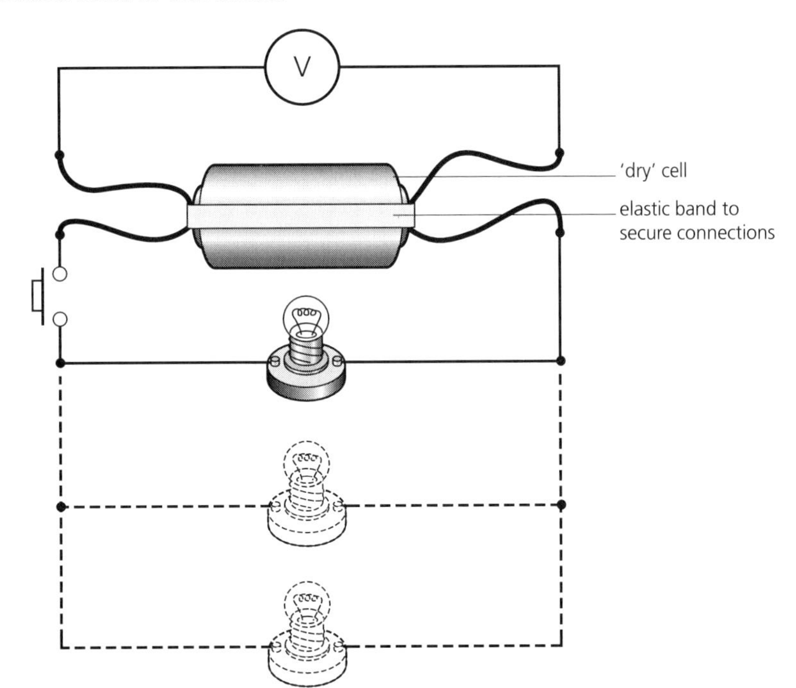

Figure 8.7

Plan

- Build the circuit of the cell and the voltmeter, and adjust the voltmeter range to give a precise reading of the 'voltage' of the cell. Record the reading with no current. This is the emf of the cell.
- Connect a lamp and a switch, and close the switch so that current flows through the resistance of the lamp (see Figure 8.7). Record the voltmeter reading again. This is the pd across the lamp and the cell.
- Add lamps one by one in parallel (this will increase the current from the cell). Record the pd each time.

Analysis

1. Record your readings systematically.
2. Write a simple qualitative conclusion about how the 'drop' in voltage depends on the size of the current.

Computer simulation InternalR

Aim **To measure the internal resistance of a battery using only a voltmeter.**

Apparatus ■ computer running the program 'InternalR' from the CD

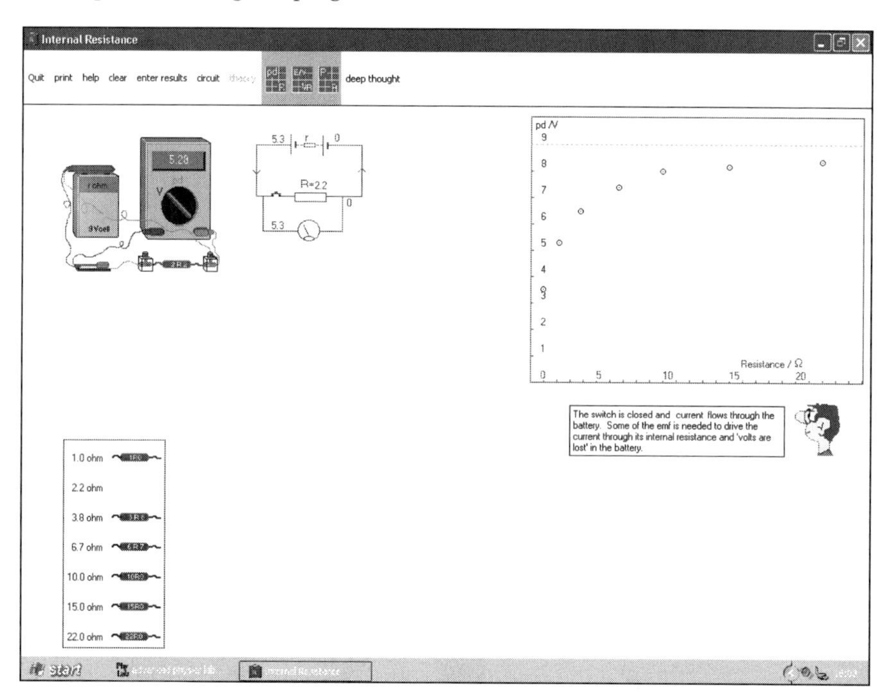

Plan

- Click 'enter results' and record the emf E of the battery as shown by the open circuit. Then click on each resistor in turn to put it into the circuit.
- Press the switch and record the pd V for each resistor in the readings box (or use the 'Auto' button to record readings automatically).
- Enter the readings into the results table by clicking 'Enter data in table'.
- Look at each of the graphs of your readings by clicking on the graph buttons at the top of the screen.

Analysis

1. Print a copy of the results. (The 'print' button prints the results, the circuit diagram, the theory and the graph currently on the screen.)
2. Use the values and the graphs to look for a pattern between the pd V and the load resistance R.
3. The graph of E/V against $1/R$ uses all of the readings to give an average value of the internal resistance of the battery.
4. The P against R graph shows how the output power P from the circuit varies with the load resistance R.
5. Write a conclusion stating:

   - your value of the internal resistance;
   - how the pd changes with the size of the load resistor;
   - how the power delivered by the battery depends on the load resistor.

Experiment: Measuring the internal resistance of a 'dry' cell

Apparatus As for the Preliminary work (p. 83), plus:

- ammeter (or multimeter set on dc amps)
- extra lamps and links

Plan

- Build a circuit with the cell, switch, ammeter and one lamp in series. Add the voltmeter across the cell.
- With one lamp in the circuit, switch on briefly and record the pd (V) across the cell and the current (I) through it.
- Add lamps in parallel one by one. For each addition record V and I.
- Repeat but adding lamps in series.
- Measure the emf (E) of the cell at intervals because, if it is a cell at the end of its life, its emf may change.

Skill level (Implementing)

> A: I built the circuit with lamp, switch, ammeter and voltmeter without help; I measured and recorded emf and pd for up to 3 lamps in parallel. I measured the pd and current for up to 3 lamps in series, recording the readings in a table. I caused no damage to the meters or lamps and did not connect the ammeter without a resistance. I switched off between readings and worked with confidence and skill.
>
> All but one of the above = B; all but two = C; all but three = D; all but four = E.

Analysis

1 Record your readings of pd and current in order of increasing current.

2 Leave an extra column in your table for a calculated value of internal resistance (r).

For the whole circuit:

$$E = V + Ir$$

$$r = \frac{(E - V)}{I}$$

3 Plot a large-scale graph of pd (V) against current (I). Draw the best-fit straight line through the points.

The equation of the line is

$$V = -Ir + E$$

Comparing this with $y = mx + c$, we can see that the gradient of the line is $-r$.

4 Measure the gradient to get a value of the internal resistance of the cell. Quote this to an appropriate number of significant figures.

5 Describe how the pd of the cell drops as it delivers more current.

Sample readings emf at start = 1.563 V
emf at end = 1.556 V

lamps	current /A	pd /V	r /Ω
4 in series	0.092	1.503	0.62
3 in series	0.098	1.498	0.63
2 in series	0.101	1.488	0.71
1 only	0.131	1.455	0.80
2 in parallel	0.259	1.373	0.72
3 in parallel	0.377	1.307	0.67
4 in parallel	0.445	1.262	0.69
		average	0.69

Plot the graph suggested in the *Analysis* for these sample readings, and use it to get a value of internal resistance.

Answer $r \approx 0.61\ \Omega$; there were 3 anomalous points that were ignored.

Evaluation The overall uncertainty can be estimated from the range of values of internal resistance in the table. If you judge the internal resistance to be constant, you can average the values and work out the variation from the average. For the sample readings it comes to $0.69 \pm 0.05\ \Omega$, or about 7%.

The graphical method of calculating *r* gives a more reliable result because it shows up anomalous values that can be ignored or re-measured.

Variation of the emf during the experiment introduces an uncertainty that is difficult to evaluate. An average can be taken or the experiment repeated with a fresh cell.

The measurements have been recorded to the 3 decimal places given by the instruments, although the measurements are not as precise as this. The calculated value of *r* is given to a more realistic number of significant figures.

Improving the plan A cell has resistance because the ions that carry the charge have to pass through the electrolyte from one terminal to the other. The chemicals change with the cell's age, the size of the current and the state of charge, and so internal resistance may not be constant. If possible you should experiment with a fresh cell and use it to pass currents for only a short time, especially when these currents are large.

8.5 Potential dividers

Preliminary work: Using a rheostat to provide a variable voltage
Experiment 1: Using a potentiometer to provide a variable voltage
Experiment 2: Dividing a voltage by a calculated amount (e.g. to supply 3 V from a 9 V battery)

Preliminary work — Using a rheostat to provide a variable voltage

Apparatus

- 12 V dc voltage supply
- rheostat (2 A 6 Ω or thereabouts)
- 12 V 24 W lamp in holder (e.g. ray box lamp)
- switch

Plan The idea is to make a potentiometer or 'lamp dimmer'.

- Connect the voltage supply to the base terminals of the rheostat.
- Connect the lamp to the slider terminal and one of the base terminals.
- Check that your 'dimmer' can vary the brightness of the lamp from off to fully bright.
- Change the lamp connection to the other base terminal (see Figure 8.9) and report on the difference.

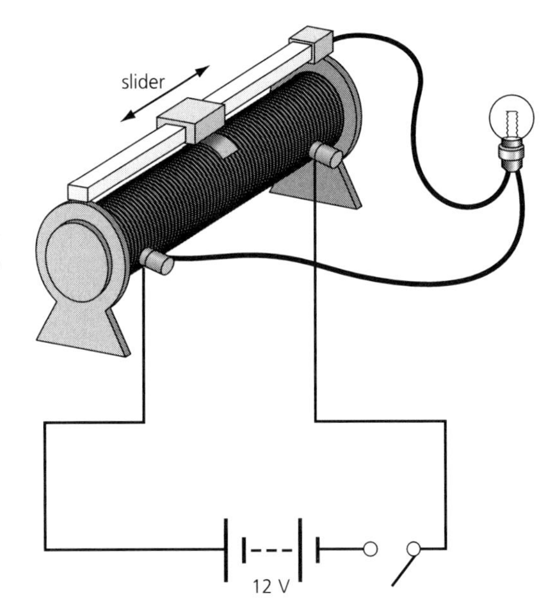

Figure 8.8

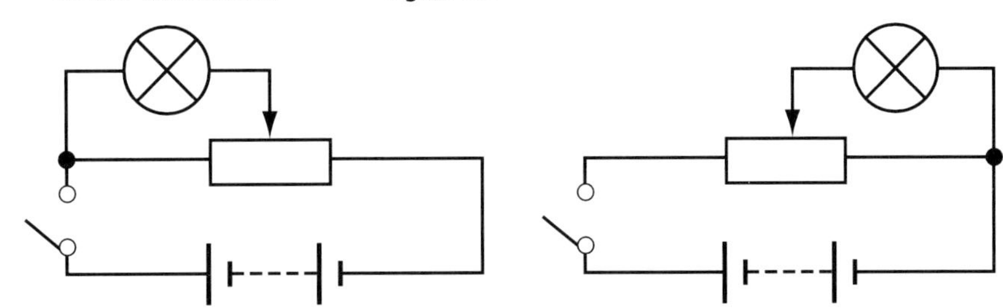

Figure 8.9

Evaluation A potentiometer gives a full range of output voltage, but its resistor gets hot and wastes energy. A high current through part of it can cause the windings to fail. Voltage control for high power or 'mains' equipment is better done using electronic circuitry.

Experiment 1 Using a potentiometer to provide a variable voltage

Apparatus

- linear potentiometer (~10 kΩ)
- logarithmic potentiometer (~10 kΩ)
- 9 V battery or dc voltage supply
- voltmeter set to measure up to 10 V
- switch

Plan

- Connect the battery or voltage supply to the outside tags of the linear potentiometer.
- Connect the voltmeter to the central tag and one of the outside tags.
- Turn the knob of the potentiometer and note how the voltage of its central tag changes.
- Repeat with the log potentiometer.

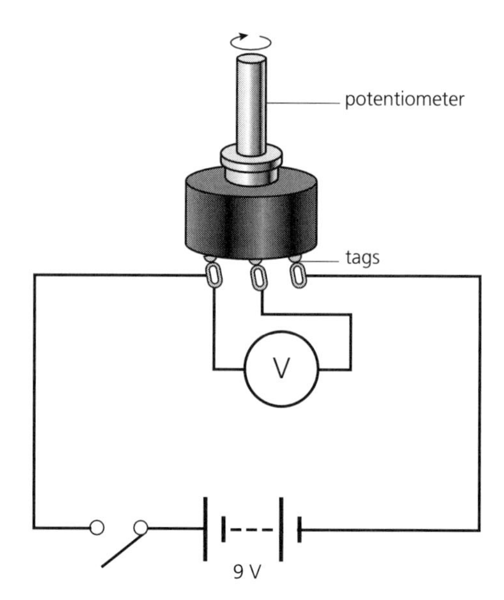

Figure 8.10

Skill level (Implementing)

> A: I soldered or attached leads to the potentiometer. I built the circuit without help. I adjusted the voltage supply and voltmeter range to ~10 V. I used the potentiometer to control the voltage from 0 V to 10 V. I did the same with the log potentiometer.
>
> All but one of the above = B; all but two = C; all but three = D; all but four = E.

Analysis

Explain the difference between the way linear and logarithmic potentiometers control voltage.

Experiment 2 Dividing a voltage by a calculated amount (e.g. to supply 3 V from a 9 V battery)

Apparatus

- 3-piece terminal block
- 4.7 kΩ, 2.2 kΩ and 1.2 kΩ resistors or similar (5% gold band quality)
- 9 V battery with battery cap and leads, or dc power supply
- connecting leads
- voltmeter set to measure up to 10 V

Plan

- Screw two of the resistors and two leads into the terminal block (see Figure 8.11).
- Connect one lead of the voltmeter to the negative terminal of the battery.
- Use the other voltmeter lead to measure the battery pd and the pd across the 'right-hand' resistor (the one connected to the battery negative).
- Put different resistors into the block and repeat the measurements.

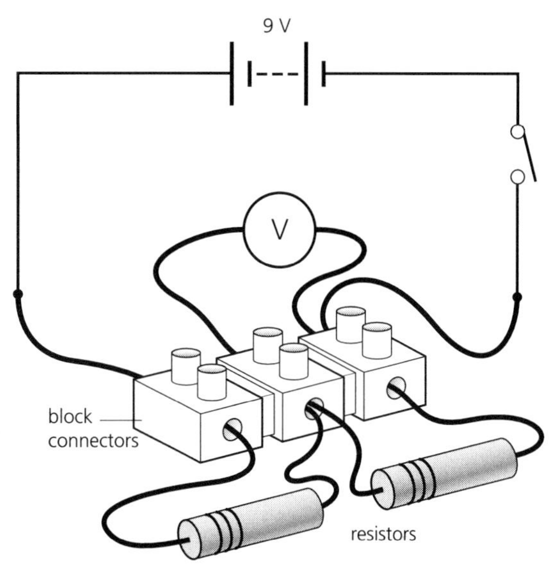

Figure 8.11

Analysis

1 Make up a table for your measurements – an example is shown in the *Sample readings* below. Include a column for the theoretical pd calculated from the values of the resistors.

$$\text{pd across } R_2 = \frac{(\text{battery pd}) \times R_2}{(R_1 + R_2)}$$

2 Write a conclusion about this way of dividing voltage.

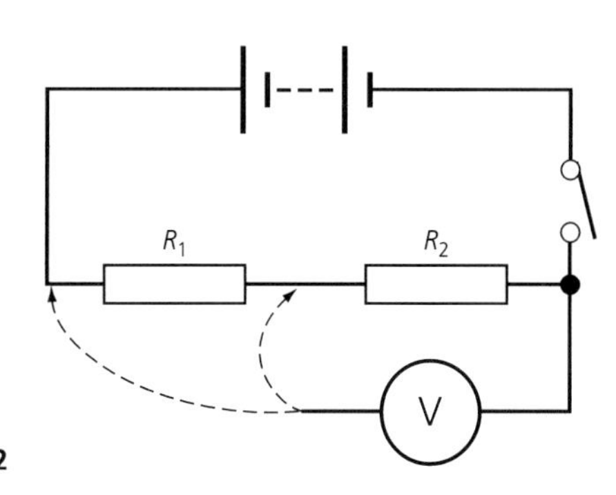

Figure 8.12

Sample readings

supply pd /V	R_1 /kΩ	R_2 /kΩ	measured pd across R_2 /V	calculated pd /V = $9.0 \times R_2/(R_1 + R_2)$
9.0	1.2	4.7	7.22	7.17
9.0	1.2	2.2	5.82	5.82
9.0	4.7	2.2	2.81	2.87

Evaluation

The difference between the measured and calculated values is mostly due to the uncertainty in the values of the resistors. The calibration and measurement uncertainties of the voltmeter contribute only a small amount to the difference.

Improving the plan

- Use better quality resistors (1% brown band quality).

8.6 Sensors

Investigation 1: Using temperature to control voltage
Investigation 2: Using light to control voltage

Investigation 1 Using temperature to control voltage

Apparatus

- 3-piece terminal block
- 9 V battery with battery cap and leads, or dc power supply
- insulated copper connecting wire
- voltmeter set to measure up to 10 V
- thermistor (resistance about 1 kΩ) with long leads
- 1 kΩ resistor
- switch
- electric kettle, or beaker of water, Bunsen, tripod and gauze
- electronic or mercury thermometer
- safety glasses (for protection from steam and hot water)

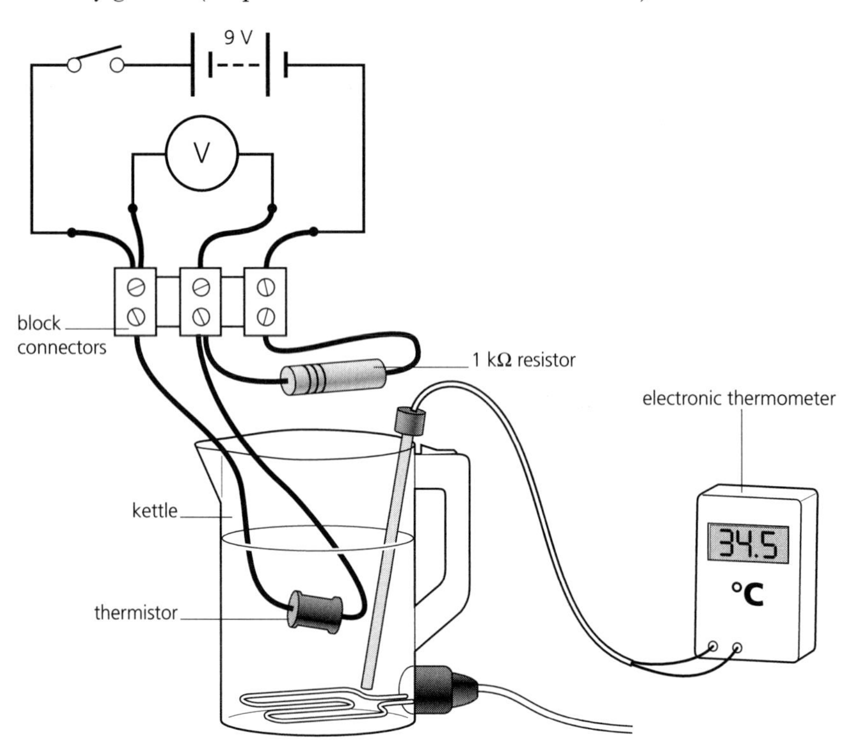

Figure 8.13

Plan

- Connect the battery, the 1 kΩ resistor and the thermistor to the terminal block (see Figure 8.13).
- Connect the voltmeter across the thermistor.
- Measure the pd across the thermistor when it is in the air and when it is in water.
- Obtain readings of pd and temperature by heating or cooling the water.

Analysis

1 Record your readings in a table, in order of increasing temperature.
2 Plot a calibration graph from which temperature can be read from voltage values.
3 Conclude whether this arrangement would make a useful thermometer.

Sample readings

Supply pd = 6.96 V
Resistor = 2.2 kV

temperature /°C	26.8	32.3	44.0	53.3	63.6	73.9	82.2	90.6	100.9
pd /V	1.04	0.91	0.63	0.49	0.38	0.29	0.23	0.19	0.14

Plot a graph of pd against temperature for the sample readings. Measure the gradients of the curve at 30 °C and 90 °C. These are a measure of the sensitivity of the circuitry to temperature change at the two temperatures.

Answers ~24 mV/°C; ~4 mV/°C.

Extension

To make this a full investigation, use a datalogger to record the temperature and voltage readings. You could use different resistors to find the one that gives the greatest change of pd with temperature.

Investigation 2 Using light to control voltage

Apparatus

Circuit as for Investigation 1, plus:

- light dependent resistor (LDR) (e.g. ORP12)
- 10 kΩ variable resistor
- lamp or torch
- clamp and stand
- black card
- aluminium foil
- pin

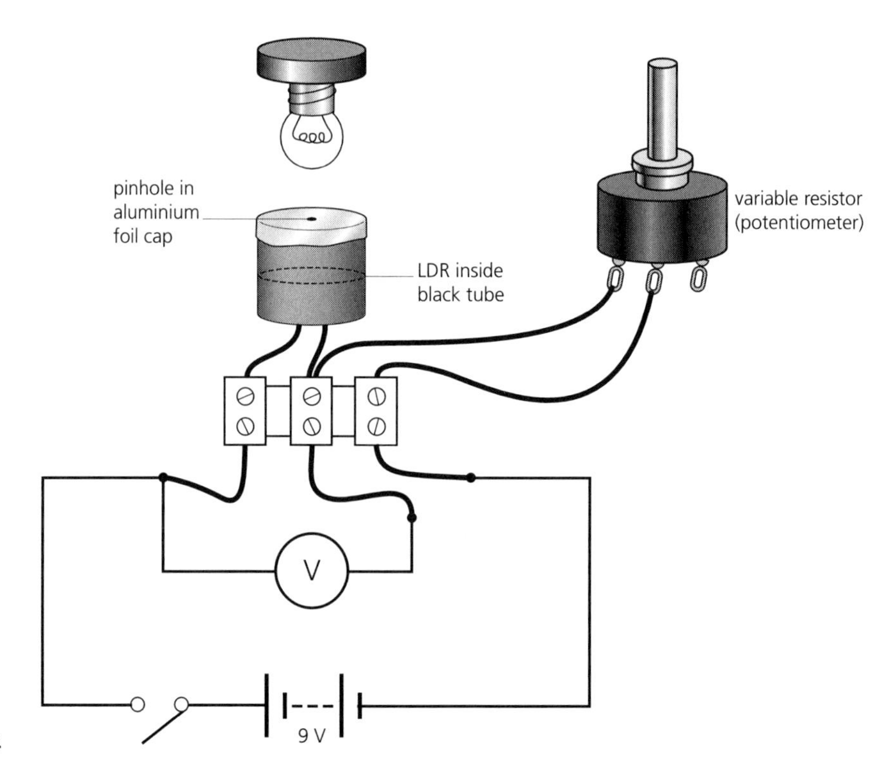

Figure 8.14

Plan

- Replace the resistor in the circuit with the variable resistor (middle and end tags).
- Replace the thermistor with the LDR.
- Connect the voltmeter to measure the pd across the LDR.
- Shine a light on the LDR and adjust the variable resistor to give a measurable voltage reading. Check that the voltmeter reading varies with light level.
- Make a black tube with an aluminium foil cap to cover the LDR. Fix a constant light source above the cell and make a pinhole in the centre of the cap to let in the light. Increase the light level in equal steps by making more pinholes. Measure the pd each time.

Analysis

1. Tabulate your results.
2. Plot a graph of pd versus number of holes to show the characteristics of your 'light meter'.
3. You could also plot a log graph (ln (pd) against ln (number of holes)) to see if this indicates an empirical equation connecting pd and number of holes (see the *Sample readings* below).

Sample readings

(a) Battery emf = 2.9 V; resistor = 1.0 kΩ

no. of holes	1	2	3	4	5	6	7	8	9	10	11	12	13	14	15
pd /V	2.50	2.33	2.18	1.98	1.89	1.89	1.75	1.73	1.64	1.55	1.50	1.43	1.42	1.38	1.33

Plot the graphs suggested in the *Analysis* for these readings.

(b) Readings obtained by measuring the resistance of an LDR directly with a multimeter on its 'Ohms' range

no. of holes	1	2	3	4	5	6	7	8	9	10	11	12	13	14	15
resistance /kΩ	26.9	18.1	14.2	11.5	9.4	8.4	7.9	7.1	6.5	5.9	5.2	5.0	4.8	4.7	4.3

It is suggested that the resistance R and the number of holes N are connected by a power law of the form $R = aN^b$, where a and b are numerical constants. Plot a graph of $\ln R$ against $\ln N$ to see if this is true (see p. 75), and use it to obtain the values of a and b.

Answer $R = 29.4\,N^{0.70}$.

Extension To make this a full investigation, find for yourself an equation that allows you to calculate the resistance of the LDR for a given number of pinholes in its cover.

Skill level (Analysing)

A: I put my readings into a properly headed table. I plotted a graph of resistance (or pd) against the number of holes. I plotted a log/log graph of resistance and number of holes. I measured the intercept and gradient. I concluded with an empirical equation for resistance.

All but one of the above = B; all but two = C; all but three = D; all but four = E.

9 Capacitance

9.1 Charging and discharging a capacitor

Preliminary work: Using a battery and a light emitting diode (LED)
Experiment: Measuring the rate of discharge of a capacitor through a resistor
Investigation: Measuring capacitance from a discharge curve

Preliminary work — Using a battery and a light emitting diode (LED)

Apparatus

- small 9 V battery
- LED with protective resistor (e.g. 33 Ω)
- 100 μF capacitor
- connecting leads

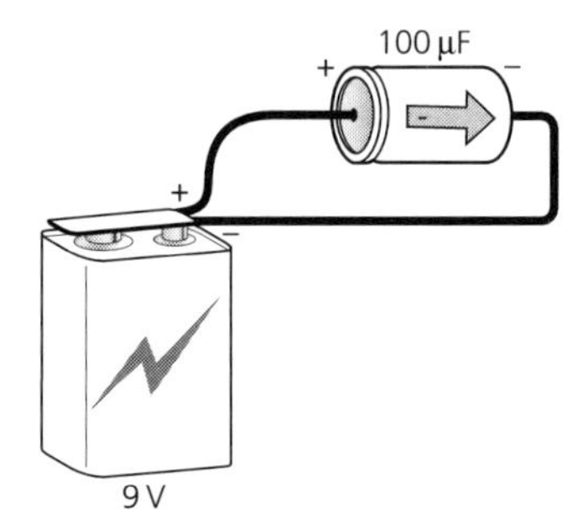

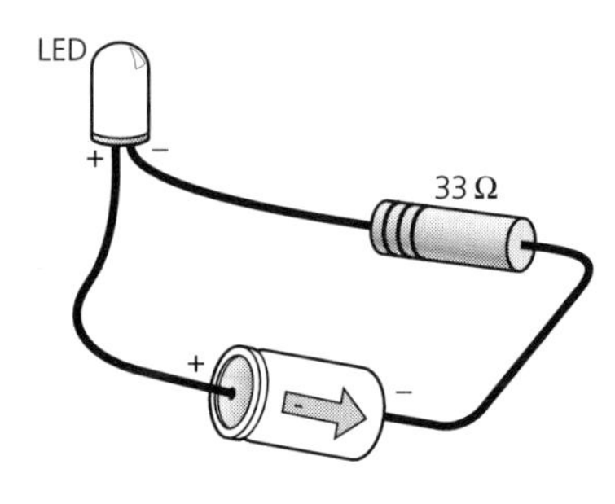

Figure 9.1

Plan The idea is to charge the capacitor using the battery and then discharge it through the LED.

- Charge the capacitor by connecting it to the battery terminals. (Make sure the + end of the capacitor goes to the + terminal of the battery.)
- Connect the charged capacitor to the LED and resistor. (Make sure the LED is connected correctly, as it is a diode and only conducts in one direction. See Figure 9.1.) Look at the LED end-on to see if it lights as the capacitor discharges.

Analysis Report on how long it takes the capacitor to discharge through the LED, i.e. for how long it can light the LED.

Experiment — Measuring the rate of discharge of a capacitor through a resistor

Apparatus

- 100 μF or 220 μF capacitor
- 330 kΩ resistor
- digital voltmeter (set to measure up to 20 V)
- dc variable power supply
- connecting blocks and leads
- switch
- stop clock or watch

Plan The idea is to measure the decrease in pd across the capacitor as it discharges.

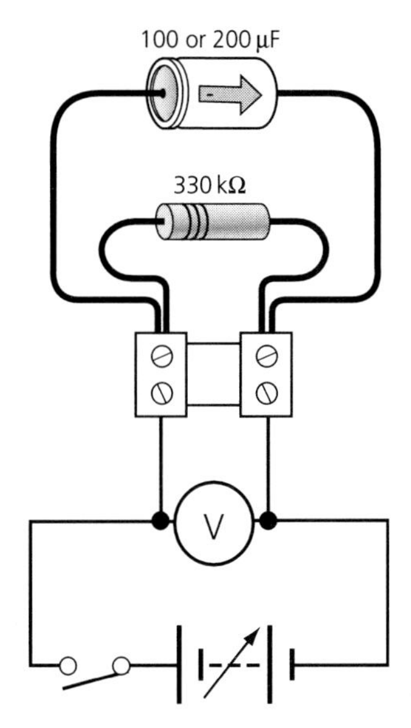

Figure 9.2

- Build a circuit that lets you charge the capacitor and then discharge it through the resistor. One arrangement is shown in Figure 9.2. Instead of using a switch you could push one lead in and out of one of the power supply sockets.
- Before you connect the capacitor, set the power supply to a safe voltage that is less than the maximum printed on the capacitor.
- Do a test run. Charge the capacitor by plugging into the power supply. The voltmeter will then show the supply pd. Start the discharge of the capacitor by unplugging the lead from the power supply. Get a rough idea of how long it takes the pd to drop to a small value.
- Decide how often you are going to take readings and then re-charge the capacitor.
- Start the stop clock as you disconnect the power supply and record pd and time until you have enough readings to represent the discharge.

Skill level (Implementing)

A: I built the circuit myself without help. I arranged the wiring neatly with no short circuits and with the supply pd less than the capacitor rating. I connected the capacitor to the dc supply with the correct polarity. I obtained enough readings to plot a smooth discharge curve. I took readings more frequently when the discharge was rapid and used the best range on the voltmeter for precise measurements.

All but one of the above = B; all but two = C; all but three = D; all but four = E.

Analysis

1. Put the readings into a clearly headed table. Record the stop watch reading as it is seen in minutes and seconds. Include a column for calculating the time in seconds. Give these values to the nearest second because, although many stop watches read to 2 decimal places, reaction time and reading uncertainties will make the time accurate to only about 1 second.
2. Plot a graph of pd against time and draw a best-fit line through the points.
3. Write a conclusion describing how the pd decreases with time.

Sample readings $C = 220\ \mu F$; $R = 330\ k\Omega$

time reading (min:sec)	0:00	0:10	0:20	0:30	0:40	0:50	1:00	1:15	1:30	1:45	2:00	2:30	3:00
time /s	0.0	10	20	30	40	50	60	75	90	105	120	150	180
pd /V	10.0	8.70	7.57	6.62	5.70	5.00	4.36	3.55	2.91	2.35	1.91	1.27	0.84

Plot the graph suggested in the *Analysis* for the sample readings.

Evaluation The voltmeter and stop watch readings are changing as they are read. Taking 'moving readings' is an imprecise experimental technique. The time value may lie half a second either side of the timing mark, leading to an uncertainty of about 5%. During this 1 second interval the pd will have changed by about 0.5 V at the beginning of the discharge, giving an uncertainty of roughly 5%. The calibration uncertainties of the voltmeter and stop watch will be small in comparison with this.

Improving the plan

- Devise a count-down technique that one observer uses to give warning of the approaching time interval. Use a second observer to read off the pd.
- Use the 'lap time' facility of a stop watch. Press the 'lap time' button the instant you read the voltmeter. Record the pd and time. The watch will carry on timing while you do this and after a pause will return to the running display.
- Use a datalogger to obtain the readings of pd and time automatically. The time taken to make the reading will be very short and the uncertainty in the pd and the time will be set by the electronics of the unit.

Investigation: Measuring capacitance from a discharge curve

Apparatus As for the previous experiment, plus:

- datalogger, if available, instead of voltmeter and stop watch
- range of resistors (e.g. 100 kΩ, 220 kΩ, 270 kΩ, 330 kΩ, 470 kΩ if using a stop watch, or 10 times less if using a datalogger)
- multimeter with resistance range

Plan The method of the previous experiment is used to obtain discharge readings for 5 different resistor values.

- Put the meter on its 'ohms' range and measure the resistance of each resistor (see p. 68).
- Build the circuit as in the previous experiment (Figure 9.2) with one of the resistors.
- If you have a datalogger, arrange it to measure the pd across the resistor. Choose a sampling rate about 1/10th of the time constant of the circuit ($C \times R$). Start the datalogger and let the capacitor discharge for a time equal to about 3 time constants.
- Use the software to print out the readings and a graph of pd against time.
- Repeat for the other resistors.

Analysis

1 For each resistor value record the readings – time, in minutes and seconds, from the stop watch and pd from the voltmeter – in a suitable table.

The theoretical equation of the discharge curve is:

$$V = V_0 e^{-t/CR}$$

V_0 is the pd at time 0.
C is the capacitance.
R is the resistance.
V is the pd at time t after the start.

When $t = CR$:

$V = V_0 \mathrm{e}^{-1} = 0.37V_0$

So the time taken for the pd to fall to 0.37 of its initial value = CR.

2 Plot the pd/time graphs for each resistor.
3 Calculate 37% of the initial pd. Read off the time from the graph when the pd has reached this value. This method can be repeated for any starting value of V (see Figure 9.3). This time interval = CR so, to find C, divide it by R.
4 Use the discharge curves for the different resistors to get an average value of capacitance.

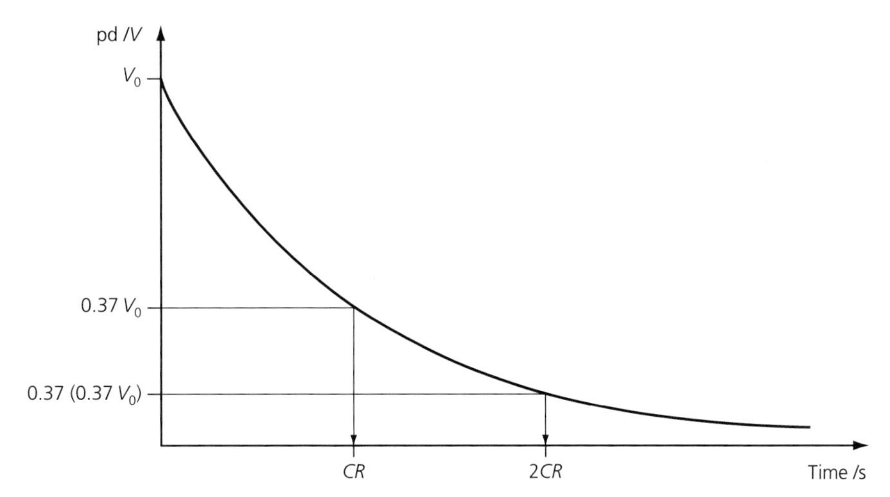

Figure 9.3

Alternative analysis

1 For each resistor value record the readings – time, in minutes and seconds, from the stop watch and pd from the voltmeter – in a suitable table.

Theory predicts that the discharge of the capacitor is exponential and the pd varies with time according to the equation:

$$V = V_0 \mathrm{e}^{-t/CR}$$
$$V_0/V = \mathrm{e}^{t/CR}$$

Taking logs to base e, this becomes:

$\ln (V_0/V) = t/CR$

A graph of $\ln (V_0/V)$ against t should therefore be a straight line through the origin with a gradient of $1/CR$.

2 Plot the log graph for each resistor and obtain an average value of C from the gradients of the best-fit lines through the origin.

Skill level (Analysing)

A: I plotted 5 precision graphs of *V* against *t* with large scales, labelled axes, units, fine points, and smooth best-fit curves. I read off the 5 values of the time constant and used them to calculate an average value for capacitance. I wrote a conclusion about the time constant and noted a similarity with radioactive discharge. I also plotted a log graph to get a straight line. I read the time constant and capacitance from the log graph.

All but one of the above = B; all but two = C; all but three = D; all but four = E.

Sample readings

Using a datalogger, sampling every 500 ms

time /s	0.0	0.5	1.0	1.5	2.0	2.5	3.0	3.5	4.0	4.5	5.0	5.5	6.0	6.5	7.0
pd /V	10.7	9.2	7.4	6.0	4.8	3.9	3.2	2.5	2.1	1.6	1.3	1.0	0.8	0.6	0.5
ln (V_0/V)	0.0	0.2	0.4	0.6	0.8	1.0	1.2	1.5	1.6	1.9	2.1	2.4	2.6	2.9	3.1

$V_0 = 10.7\,V$; $R = 7450\,\Omega$

Plot a log graph for these readings, as suggested in the *Alternative analysis*, and obtain a value for the capacitance of the capacitor.

Answer ~320 μF.

Evaluation Reading the time constant (*CR*) from one point on a curved graph (as in the first analysis) is not a precise technique. It uses only the readings in the neighbourhood of that point, and the curve may be distorted over that region. Plotting a log graph as in the second analysis is much better because the best-fit line is chosen with reference to all of the points. Even so, a range of gradients is possible and the uncertainty can be estimated from this variation (about 6% for the sample readings).

The resistance of the voltmeter is in parallel with the discharge resistor and will reduce the total resistance. With large resistor values, the uncertainty in *R* will be significant. A 1 MΩ voltmeter in parallel with a 470 kΩ discharge resistor, for example, gives a total resistance of 320 kΩ (an uncertainty of 32% if you ignore the effect of the voltmeter).

Improving the plan

- Use a smaller discharge resistor, e.g. 10 kΩ. 1 MΩ in parallel with 10 kΩ gives a combined resistance of 9.9 kΩ – an uncertainty of only about 1%. The capacitor will discharge quickly, so you will need a datalogger to record the readings. (The datalogger will probably have about the same resistance as the voltmeter.)
- Stop recording the pd when its value becomes small (e.g. less than 1.00 V) because the fluctuations of the readings will introduce a large uncertainty.

Note The capacitance value obtained from the graphs could be very different from the value printed on the component. The printed value is typically accurate to only 20% and may only apply when using the correct working voltage.

Extension To make this a full investigation, find out if the capacitance of a capacitor varies with the voltage across it.

9.2 Capacitance measurements

Computer simulation: Capacitance
Preliminary work: Rapidly charging and discharging a capacitor
Experiment: Finding out if pd and charge are proportional for a capacitor

Computer simulation

Capacitance

Aim **To understand the idea of capacitance and the factors that affect its value.**

Apparatus

- computer running the program 'Capacitance' from the CD

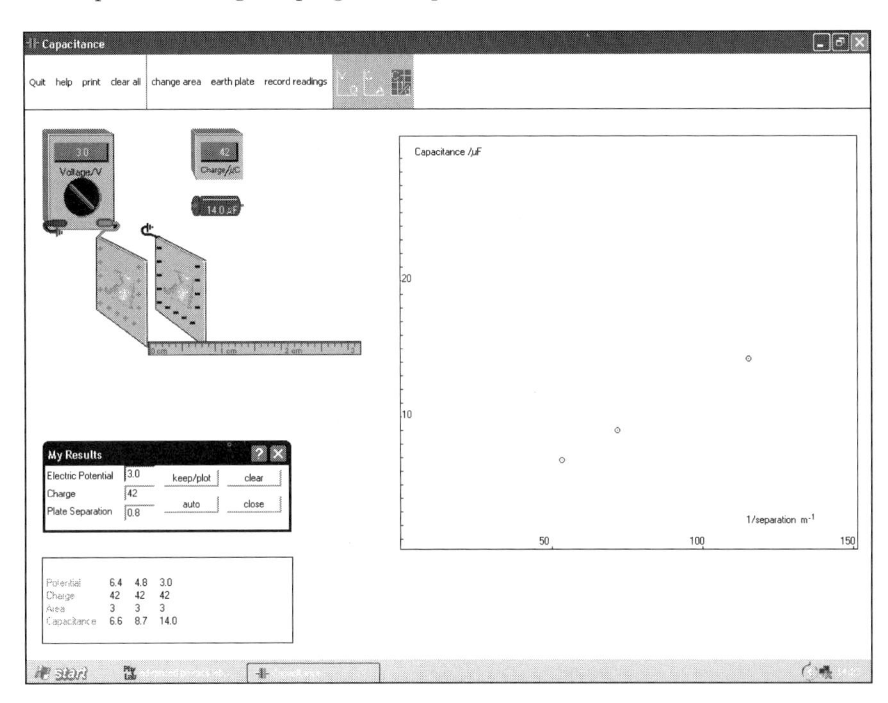

Plan Use this imaginary experiment to put electric charge on an insulated metal plate. As the charge rises, so does the potential of the plate.

- Click on the plate to add positive charge in small amounts. For each addition, record charge and potential in the results table. Plot the V/Q graph to show that charge and potential are proportional.
- Repeat with a larger plate area. You will see that the larger plate can store the same charge at lower potential. It is said to have a larger capacitance.
- Introduce an earthed metal plate, as shown above. This acquires an induced negative charge that reduces the potential of the charged plate. Since the charge remains the same, the earthed plate increases the capacitance. Two plates like this form the basic structure of a parallel plate capacitor.
- Work carefully through the four experiments that are suggested in the on-screen 'help' notes, taking readings and plotting graphs.

Analysis

1. Print out results and graphs from each of the four experiments.
2. Write a concise conclusion to go with them.

Preliminary work

Rapidly charging and discharging a capacitor

Apparatus

- 470 μF capacitor
- variable dc power supply
- analogue ammeter (or analogue multimeter)
- 2-way relay or reed relay (DPST)
- low voltage ac power supply (50 Hz)

Plan The idea is to see if rapid movement of charge from a capacitor produces a current.

- Connect the relay so that one contact charges the capacitor from the power supply and the other discharges it through the ammeter (see Figure 9.4).
- Adjust the output of the ac power supply to the voltage required by the relay coil. Connect the coil of the relay to the supply, switch on and check that it buzzes.
- Adjust the range of the ammeter until it registers a current.
- Note how the ammeter registers the rapid pulses of charge.

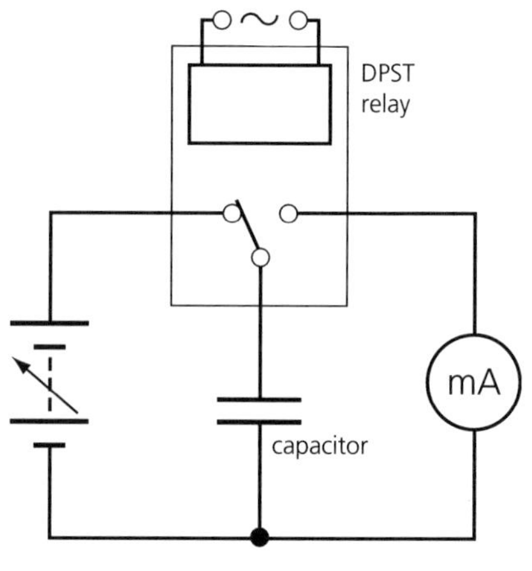

Figure 9.4

Analysis Electric current is the flow of charge. The capacitor gains charge from the dc power supply and releases it through the ammeter. This happens 100 times a second and so charge flows through the ammeter in rapid bursts. (There are 50 voltage cycles a second from the ac supply but, unless there is a diode in the coil circuit, both the positive and negative parts of the voltage cycle will switch the relay.)

Report on whether your ammeter shows these bursts as a steady current, or whether any vibration of its pointer is visible.

Experiment

Finding out if pd and charge are proportional for a capacitor

Apparatus As for the Preliminary work above, plus:

- voltmeter

Plan

- Build and adjust the equipment as in the Preliminary work.
- Connect the voltmeter to measure the pd across the dc supply (V).
- Record the switching frequency (f). This will probably be double the mains frequency: 2×50 Hz in the UK.
- For 8 different values of pd V, measure the discharge current (I).

Skill level (Implementing)

A: I built the circuit neatly and made it work without help. I selected a suitable range on the ammeter to register the current. I changed the pd and got 8 values of pd and current. I chose the correct voltage to work the relay and kept the capacitor pd below its maximum. I did not damage the reed switch or relay and capacitor, or overload the ammeter.
All but one of the above = B; all but two = C; all but three = D; all but four = E.

Analysis

1 Present your readings in a way that clearly identifies the quantities and their units. They should be given to the number of decimal places shown by the instruments.

The theory is that:

$$\text{average current} = \text{charge} \times \text{frequency}$$
$$I = Qf$$
$$Q = I/f$$
$$\text{and capacitance} = \text{charge/pd}$$
$$C = Q/V = I/fV$$
$$\therefore \quad V = I/fC$$

So a graph of V against I should be a straight line if pd and charge are proportional (i.e. if the capacitance is constant). The gradient of the line will be $1/fC$, from which C can be calculated.

2 Plot the graph of V against I and use it to calculate a value for capacitance.

Sample readings

Mains frequency was used to switch the relay. A check showed that the relay switched twice a cycle, i.e. at 100 Hz.

pd /V	0.0	1.0	2.0	3.0	4.0	5.1	6.1	7.0	8.0	9.0	9.9
current /mA	0.0	4.0	10.0	12.0	18.0	24.0	28.0	30.0	32.0	38.0	42.0

Plot the graph suggested in the *Analysis* for these sample readings.
Are there any anomalous points?
Does it show that pd and charge are proportional?
Calculate capacitance from the gradient.

Answer ~42 μF.

Alternative analysis using a spreadsheet

It is a good experimental technique to put readings into a spreadsheet as you measure them. The table shows the sample readings in a spreadsheet arrangement you can use for this experiment.

	A	B	C	D	E	F	G	H	I	J	K	L	M	N
1	pd /V	0.0	1.0	2.0	3.0	4.0	5.1	6.1	7.0	8.0	9.0	9.9		
2	current /mA	0.0	4.0	10.0	12.0	18.0	24.0	28.0	30.0	32.0	38.0	42.0		
3	frequency /Hz	100	100	100	100	100	100	100	100	100	100	100		
4	capacitance /μF													

You can enter formulae to calculate:

- capacitance $= I/fV$: in cell C4 enter the formula **=C2*1000/(C1*C3)** and 'fill across' to L4
- averages: in cell M4 enter the formula **=AVERAGE(C4:L4)**
- standard deviation: in cell N4 enter the formula **=STDEVP(C4:L4)**

You can also instruct the program to plot the graph of pd and current as you go along.

Skill level (Analysing)

> A: I entered the readings into a spreadsheet. I headed the columns correctly. I used the program to plot and print a graph of current against pd. I entered formulae to calculate capacitance. I entered formulae to calculate average capacitance and standard deviation.
>
> All but one of the above = B; all but two = C; all but three = D; all but four = E.

Evaluation

Mains frequency is kept very steady and the likely variation will be less than 1%. The uncertainty in pd will be set by the accuracy of the voltmeter and may be 2%.

The biggest uncertainty will be in the current reading. Reading the scale introduces uncertainties, and the instrument is not calibrated to read bursts of charge.

An estimate of the random uncertainties can be obtained from the numerical variations of the capacitance values from the average (or the standard deviation). This is 3 μF out of 43 μF for the sample readings ($\approx$ 7%). Adding the instrument uncertainties gives an overall uncertainty of about 10%. Anomalous readings should be ignored or re-measured.

Improving the plan

- Adjust the pd to give unit current readings so that the needle of the ammeter is exactly on one of the marks of its scale. Use the mirror behind the scale to make sure there is no parallax when your eye is vertically above the needle. This avoids having to estimate readings that lie between divisions.

10 Electromagnetism

10.1 Magnetic fields generated by electric currents

Computer simulation: Bfield
Experiment 1: Plotting the magnetic field pattern of a current in a coil
Experiment 2: Measuring the strength of the magnetic field of a current in a solenoid
Investigation: Measuring the variation in the strength of the magnetic field along the axis of a coil carrying a current

Computer simulation

Bfield

Aim **To investigate the shape of magnetic fields by plotting magnetic field lines.**

Apparatus

- computer running the program 'Bfield' from the CD

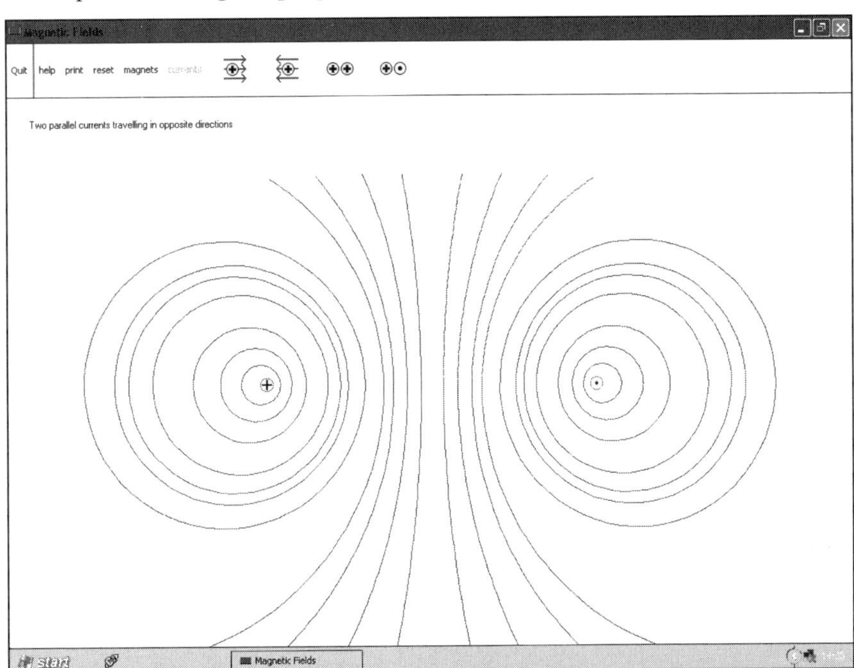

Plan

- Choose 'currents' and start with one of the options – a current in a long straight wire in an external field, or two long parallel conductors carrying currents (as shown above).
- Click near to the N pole of the external field, or near the wire carrying a current, to start a field line. Its path will then be plotted using the theoretical equations that calculate the force on a N pole in the field.

- Read the hints in the Help notes on screen and practise plotting until you have a representative picture of the field. Remember that the field lines should be evenly spaced where the field is uniform. Make sure you plot some lines close to the wire; see Figure 10.1, for example.

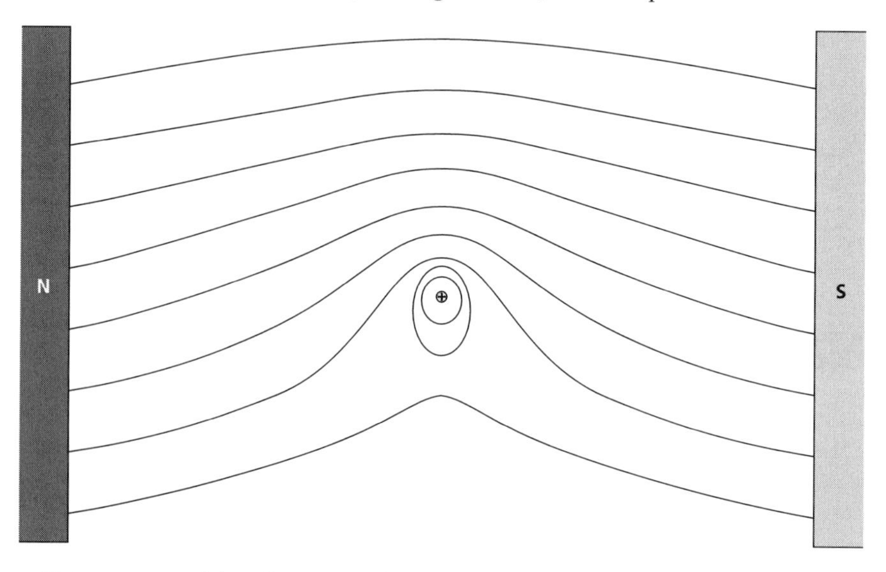

Figure 10.1

- Print a copy of the diagram and put arrows on the lines to show their direction.
- Make and print diagrams of the other field patterns generated by currents in straight wires.
- Also investigate the many shapes of magnetic fields that can be produced by magnets. Click near a N pole to plot a field line.

Analysis In all of these examples involving currents, the circular field around a current in a straight wire is distorted by a second field into a new shape. This combined field exerts a force on the wire.

On each of your diagrams draw an arrow or arrows to show the direction of the force(s) on the wire(s).

Experiment 1 Plotting the magnetic field pattern of a current in a coil

Apparatus

- large diameter coil made with wire that can take a current of several amperes (one of the Helmholtz coils of an electron beam kit is ideal)
- dc power supply and connecting leads
- cardboard box lid (an A4 paper box lid is just the right depth)
- plotting compass

Plan

- Cut the box lid so that it forms a platform through the middle and around the sides of the coil (see Figure 10.2). Cut a sheet of paper to cover the top surface of the lid.
- Use the power supply to send a current of about 1 A through the coil. Check that the magnetic field produced is strong enough to deflect the plotting compass.

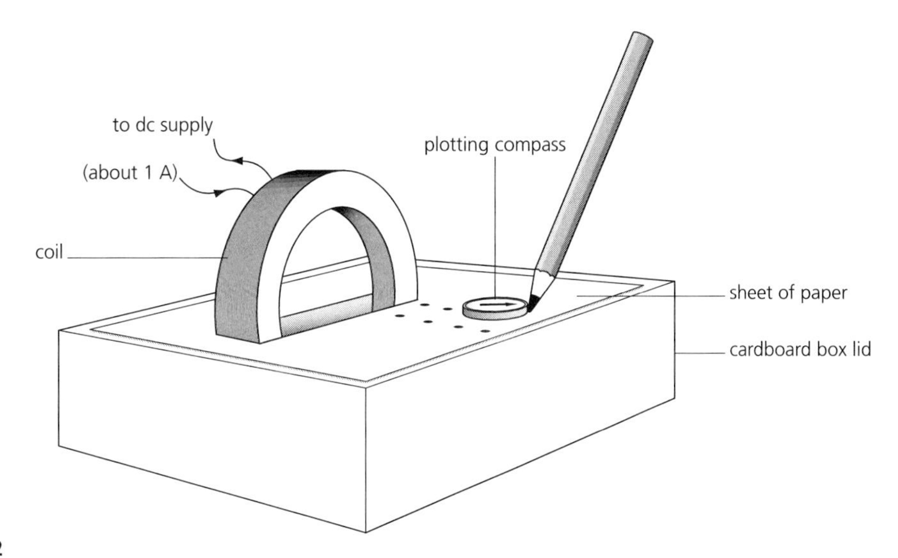

Figure 10.2

- Mark a pencil dot where you want to start a magnetic field line.
- Lay the tail of the compass needle over this dot and mark a second dot at its head. Move the compass until its tail is over the second dot and continue to mark dots until you reach the edge of the paper.
- Start from a different place and build up lines until you have enough to show the shape of the magnetic field.

Analysis

1 Join the dots to form the field lines.
2 Put arrows on the lines to show the direction in which the N pole of the compass magnet pointed.

Sample plot

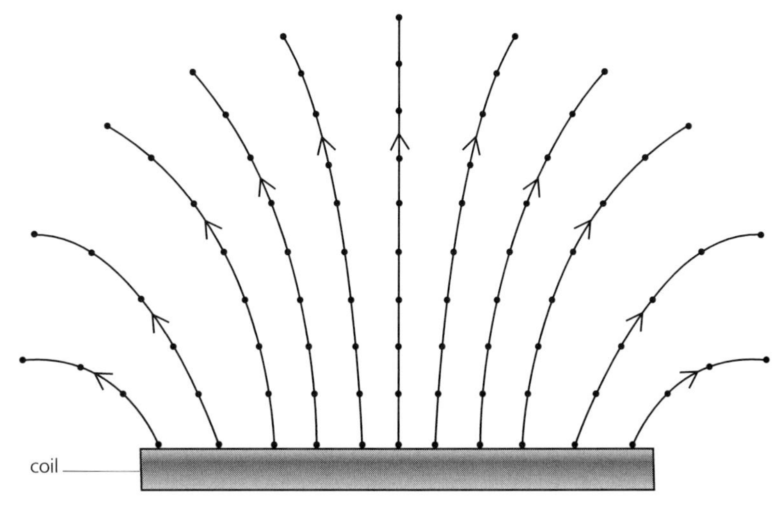

Figure 10.3

This pattern was found using plotting compasses on a platform through the middle of a Helmholtz coil carrying a current of 1 A.

Experiment 2 Measuring the strength of the magnetic field of a current in a solenoid

Apparatus

- power supply and connecting leads
- solenoid
- Hall probe and meter
- 2 half-metre rules
- clamp and stand
- masking tape

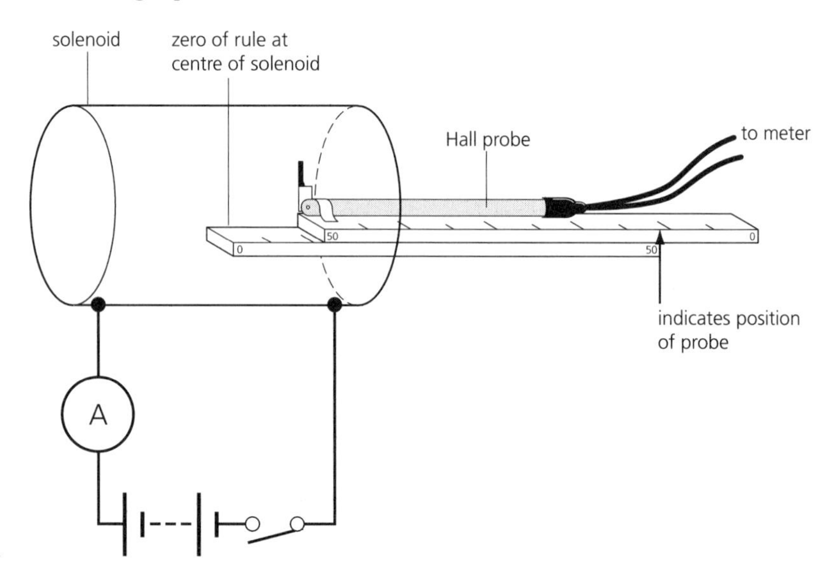

Figure 10.4

Plan

- Tape the Hall probe to a half-metre rule, with its detector at the 50 cm mark.
- Clamp the second half-metre rule inside the solenoid, with its zero exactly half-way along (see Figure 10.4).
- Switch on the current in the solenoid and adjust to its maximum safe value. (This should be given in the supplier's catalogue.)
- Connect the Hall probe to its meter and place inside the solenoid so that both rules are lined up. The detector will then be in the centre of the solenoid.
- Read the magnetic flux density in this position.
- Move the probe and measure the magnetic flux density along the axis at 1 cm intervals inside and then outside the solenoid.
- Repeat for the other end of the solenoid.

Analysis

1. Tabulate the magnetic flux density (B) and position readings (x).
2. Plot a graph of B against x to show how the magnetic flux density varies for the whole solenoid. On the x-axis mark where the ends of the solenoid would come, draw vertical lines to the top of the graph paper and cross-hatch the rectangle to represent the inside of the solenoid.
3. Write a conclusion about how magnetic flux density varies inside, at the ends of and outside the solenoid.

Sample readings The solenoid was 7 cm long.
The Hall probe gave only integer readings.

distance from centre /cm (+ to the right, − to the left)	0	1	2	3	4	5	6	7	8	9	10	−1	−2	−3	−4	−5	−6	−7	−8	−9
magnetic flux density /mT	12	12	12	8	6	4	3	1	1	0	0	12	12	8	6	4	3	2	1	0

Plot the graph suggested in the *Analysis* for these sample readings.

Investigation

Measuring the variation in strength of the magnetic field along the axis of a coil carrying a current

Apparatus

- suitable large coil (a Helmholtz coil used to deflect electron beams is ideal)
- small search coil with a large number of turns (the coil from a solenoid-operated reed switch works well, e.g. Maplin 'sealed reed relay')
- low voltage ac power supply
- digital meter that can read ac volts 0–200 mV
- metre rule
- clamps and stands

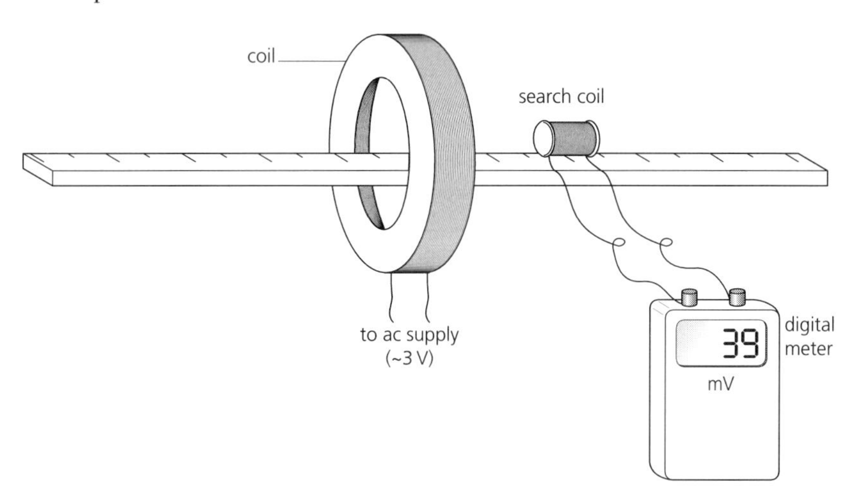

Figure 10.5

Plan

- Connect an alternating voltage of about 3 V to the coil. The current that flows will generate an alternating magnetic field around the coil.
- Clamp the metre rule along the axis of the coil with the 50 cm mark at the centre of the coil.
- Lay the search coil on the metre rule with its axis along the rule and connect it to the meter, set to ac mV.
- The alternating magnetic flux will induce an emf in the search coil that can be measured by the meter. Move the search coil along the metre rule and take readings of its position and the emf.

Analysis The emf induced by the changing magnetic flux will be proportional to its maximum magnetic flux density.

1 Record the values of the emf and the position of the search coil in a table.
2 Plot a graph that shows how the magnetic flux density varies along the axis of a coil carrying a current.

Sample readings

position /cm	40	41	42	43	44	45	46	47	48	49	50 (centre)
emf /mV	25	32	41	56	69	89	112	136	158	172	176

position /cm	51	52	53	54	55	56	57	58	59	60
emf /mV	166	149	126	103	83	65	51	39	30	24

Plot the graph suggested in the *Analysis* for these sample readings.

Extension To make this into a full investigation, you could measure the magnetic flux density along the axis of two equal coils separated by their radius (the Helmholtz position), with the current flowing (i) in the same sense and (ii) in opposite senses.

10.2 The force on a conductor in a magnetic field

Experiment: Using a current balance to measure magnetic flux density
Full investigation: Using the force on a coil carrying a current to measure magnetic flux density more precisely

Experiment — Using a current balance to measure magnetic flux density

Apparatus A simple laboratory current balance, consisting of:

- copper wire frame
- knife-edge contacts
- zero indicator
- ammeter (0–5 A)
- 2 ferrite magnets on a soft iron U-frame (e.g. from a motor kit)
- aluminium foil
- sensitive electronic balance
- variable voltage supply and connecting leads

Plan

- Balance the copper frame on the knife edges with one of the frame's ends in the magnetic field of the magnet (see Figure 10.6).
- Connect the power supply and ammeter to the knife-edge contacts so that current can flow around the copper frame.
- Switch on the current briefly to check that a force acts downwards on the part of the frame that is in the magnetic field. (If the force acts upwards, reverse the current or field direction.)
- Fix the zero indicator and carefully balance the copper frame so that it is horizontal and its pointer is at the zero mark. The frame can be made more stable by bending it downwards slightly at the half-way mark.

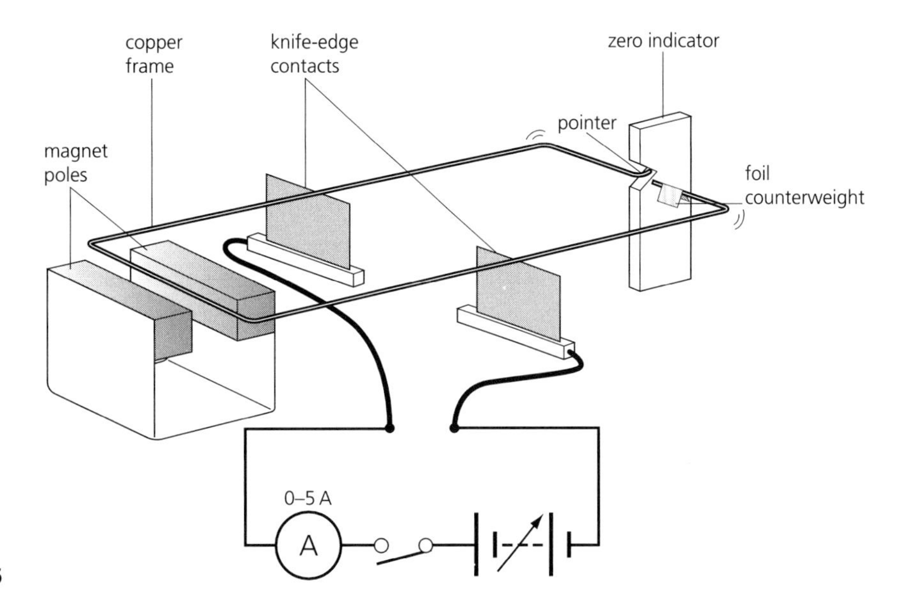

Figure 10.6

- Put a small piece of aluminium foil (about 400 mg) on the free edge of the frame and adjust the current until the frame is again level. (This is easier than trying to change the weight of the foil with a constant current.)
- Measure the length of the frame in the magnetic field (L), the current (I), and the weight of the piece of aluminium (mg).
- Repeat with pieces of aluminium foil of different sizes.

Analysis

1 Record your results in a spreadsheet with a description of each quantity and its unit.

When the frame is balanced:

force on the current in the field = weight of the piece of aluminium

$$BIL = mg$$
$$B = mg/IL$$

2 Calculate the magnetic flux density B from your results.

Sample readings

The length of the frame in the field was 5 cm.

current /A	1.0	1.4	1.8	3.8		
mass of aluminium /g	0.08	0.12	0.16	0.31		
weight of aluminium /mN	0.8	*	*	*		
magnetic flux density /mT	16	*	*	*	*	average
					*	std
					*	%

*Quantities to be calculated from the measured results.

Put these sample results into a spreadsheet and program in formulae to calculate the missing results.
Calculate the average and the standard deviation as a percentage.

Answers $B = 16$ mT; std $= 0.7$ mT $= 4\%$.

Evaluation This is not a precise experiment but it is important because it uses first principles to measure the strength of a magnetic field from the force it exerts on a current.

The chief lack of precision is in setting and reading the current needed to balance the weight of the pieces of aluminium. It is difficult to judge when the frame is exactly horizontal and the current can jump as the frame moves on the knife edges. The uncertainty in the current could be 0.2 A in 2 A (=10%).

The uncertainty in L is caused mainly by not knowing where the edges of the field are (an uncertainty of 2 mm at each end would be ~8%). The masses of the pieces of aluminium are so small that even a good electronic balance will introduce a sizeable % uncertainty (e.g. 0.02 g in 0.2 g = 10%). So the value of magnetic flux density will be uncertain to something like 30%. You may find that the standard deviation shows that your actual results are more precise than this.

Improving the plan

- Make the masses and currents as large as you can to reduce the percentage uncertainties.
- Redesign the plan to make the electromagnetic force larger. The method in the following investigation is one way of doing this.

Full investigation

Using the force on a coil carrying a current to measure magnetic flux density more precisely

Apparatus

- small rectangular coil (see below)
- clamp and stand
- variable voltage supply
- ammeter (0–5 A)
- connecting leads
- electronic balance
- ferrite magnets on an iron U-frame
- Hall probe (optional)

The rectangular coil can be constructed by winding about 10 turns of thin, insulated copper wire onto a plastic or wooden rectangular frame. The armature frames used in school motor kits are suitable. Leave long leads at each end. The magnets and U-pieces used in the same kits can be used to produce the magnetic field.

Plan

- Place the U-frame with its magnets on the electronic balance.
- Clamp the coil in position above the magnet with its bottom edge in the centre of the field (see Figure 10.7).
- Connect the power supply and ammeter to the leads of the coil.
- Note the reading on the balance.
- Adjust the emf of the supply to a few volts and switch on. Check that the balance reading changes as the force between the current and the magnetic field pushes or pulls the magnet.

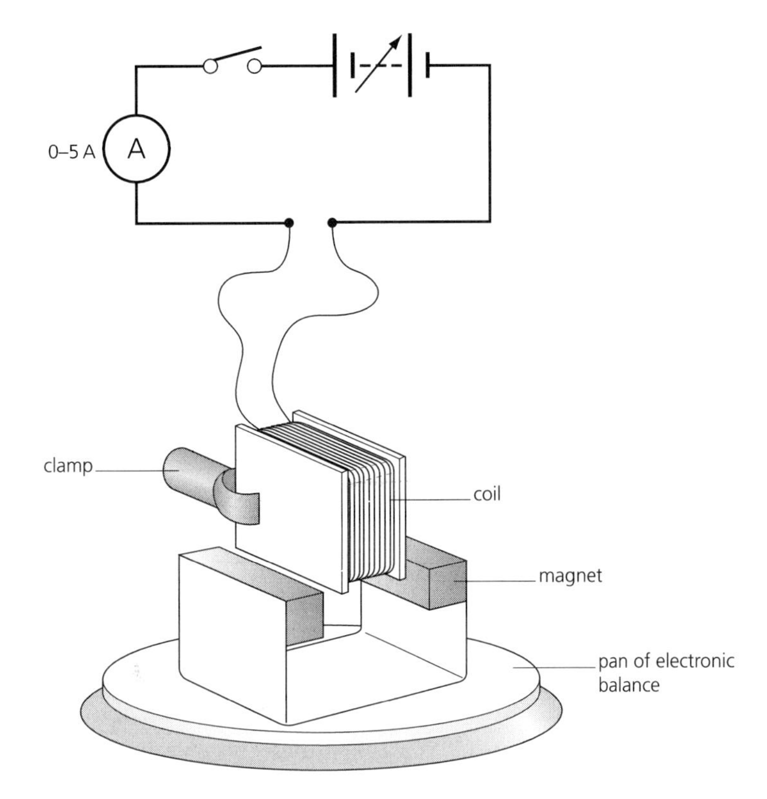

Figure 10.7

- Obtain 8 to 10 careful readings from the balance and the ammeter as the current (I) is varied up to about 3 A.
- Measure the length of the bottom edge of the coil that lies in the magnetic field (L).
- Count the number of turns on the coil (N).

Analysis

1 Put your readings into a spreadsheet, with columns for calculating the force on the coil, the magnetic flux density and the uncertainty (see the *Sample readings* overleaf).

The force on the magnets is equal to the force on the coil:

$$mg = BILN$$

so

$$B = mg/ILN$$

2 Calculate the force (mg) from the change in the balance reading; m will be shown in grams, so convert to newtons by dividing by 1000 and multiplying by g.
3 For each reading, calculate magnetic flux density from $B = mg/ILN$. These values will help you estimate the random uncertainty.
4 Also plot a graph of mg (y-axis) against I (x-axis) to help spot anomalous readings. Measure its gradient ($=BLN$) and use it to calculate a value for B.
5 Use a Hall probe if you have one, to check your value of magnetic flux density and decide which method gives the most accurate result.

Sample readings This spreadsheet was set up.

	A	B	C	D	E	F	G	H	I	J	K	L
1	balance reading /g	138.72	138.48	138.20	137.62	137.09	136.50	136.09	135.58	135.12	134.64	134.39
2	current /A	0	0.16	0.36	0.75	1.12	1.52	1.82	2.20	2.50	2.83	3.01
3	change in balance reading /g	0	0.24	0.52	1.10	1.63	2.22	2.63	3.14	3.60	4.08	4.33
4	force /mN	0	2.35	5.10	10.79	15.99	21.78	25.80	30.80	35.32	40.02	42.48
5	flux density /mT		30.4	29.3	29.7	29.5	29.6	29.3	28.9	29.2	29.2	29.2
6	numerical difference from average		1.0	0.1	0.3	0.1	0.2	0.1	0.5	0.2	0.2	0.2
7												
8	length of coil in field /cm	4.4										
9	number of turns	11										
10	average flux density /mT	29.4 ± 0.3										

To calculate the results automatically, the following formulae were put in:

in C3 **=B1-C1** and fill across to column L
in C4 **=C3*9.81** and fill across to column L
in C5 **=C4*100/(C2*B8*B9)** and fill across to column L
in B10 **=AVERAGE(C5:L5)**
in C6 **=ABS(C5-B10)** and fill across to column L

Plot the graph suggested in the *Analysis* for the sample readings, measure its gradient and calculate a value for *B*.

Answer ~30 mT.

Skill level (Analysing)

A: I entered the balance and current readings into a spreadsheet. I entered a formula to calculate force in newtons from the change in the balance reading. I used the program to plot and print a graph of force against current. I drew the best-fit line and measured the gradient of the graph. I entered a formula to calculate the average flux density.

All but one of the above = B; all but two = C; all but three = D; all but four = E.

Evaluation The readings of the balance and the ammeter are quite precise and the uncertainty in *B* is further reduced by finding the average of a large number of values. In the sample readings, the numerical difference of the readings from the average is about 0.3 out of 29.4, which is an uncertainty of only about 1%. (The standard deviation is 0.4 mT and can be used instead.)

Flaws in the design of the experiment will reduce the accuracy of the final result as follows.

- The magnetic field does not have a sharp edge and so it is impossible to know the exact length of the conductor in the field.
- The field will not be uniform and so the value for flux density will be an average. Also, the coil may move in the field.
- The wires at the top of the coil will be in part of the field and will exert an unknown force on the magnet.

Improving the plan

- Use a long frame to construct the coil, so that the top edge is as far as possible from the magnetic field.
- Use a frame that is narrower than the magnetic field so you know it is completely within the field. The length can then be measured with greater certainty.

10.3 The deflection of an electron beam by a magnetic field

Experiment: Measuring the speed of electrons in a beam by using a magnetic field to deflect the beam and recording the path

Experiment: Measuring the speed of electrons in a beam by using a magnetic field to deflect the beam and recording the path

Apparatus

- deflection tube containing deflecting plates and a screen marked in centimetre squares
- stand for the tube
- power supply for the tube (EHT and 6 V ac)
- connecting leads
- bar magnet
- Helmholtz coils
- low voltage dc power supply for the coils
- ammeter (0–1 A)

Note There are several different ways of using the Teltron 'deflection tube' illustrated in Figure 10.8, and several alternative items of apparatus. For example, the speed of an electron beam can be found using a 'fine-beam tube', in which the beam is made visible by the presence of a low-pressure gas filling.

Some of these alternatives use an HT supply (0–300 V dc) instead of or in addition to the EHT supply. A typical educational HT supply is potentially lethal because it can supply sufficient current to be fatal and it **must therefore be subject to strict supervision** when used by students. It must always be used with shrouded 4 mm plugs and the supply must always be off when changing connections.

An EHT supply is much safer, in spite of the higher voltage, because its output is limited to a safe current.

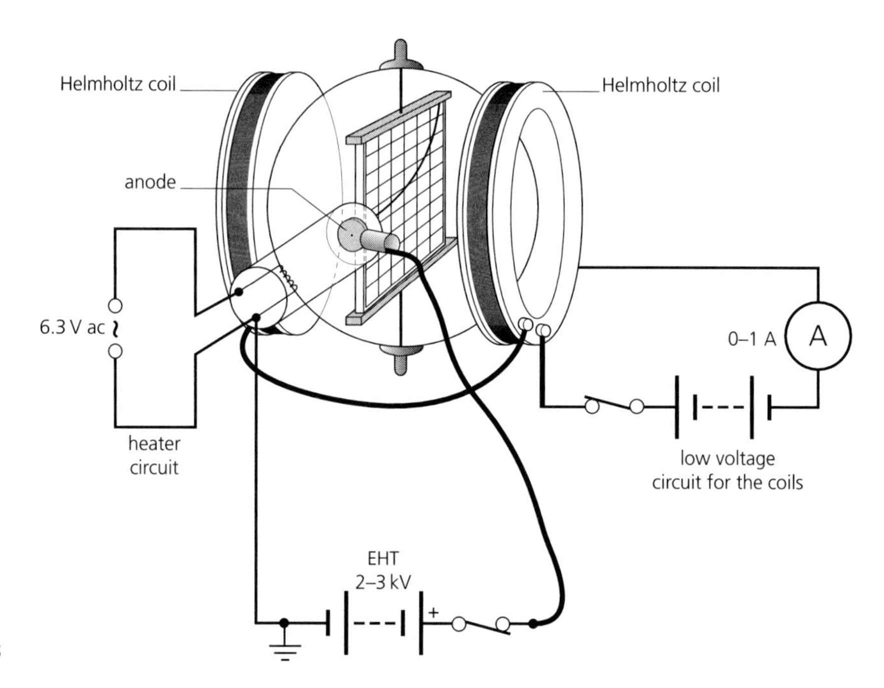

Figure 10.8

Plan

- Set up the tube in a darkened room.
- Connect the 6 V ac supply to the tube's heater coils.
- Connect the earthed negative terminal of the EHT supply to a heater connection and the positive to the anode (see Figure 10.8).
- Switch on, turn up the EHT supply to 2 or 3 kV and check you can see the fluorescent line produced on the screen by the electron beam.
- Bring up the N pole of the magnet to the side of the beam and observe the direction and shape of the deflected beam.
- Switch off the EHT supply and the heater supply before setting up the Helmholtz coils.
- Fit the Helmholtz coils either side of the tube and connect them in series with the ammeter and the low voltage dc power supply (see Figure 10.8).
- Adjust the current in the coils to give a suitable deflection on the screen and record x and y co-ordinates of the line from the scale on the screen.
- Read the current (I), the number of turns on one of the coils (N) and the mean radius of one of the coils (r). (This will also be the distance between them.)

Analysis

1 Draw a diagram of the direction of the electron beam, the magnetic field from the bar magnet and the resultant force on the electrons. Explain how the directions fit in with Fleming's 'left hand rule'.

2 Record the readings of I, N, r for the Helmholtz coils.
3 Find the radius (R) of the circular arc of the deflected beam by plotting its co-ordinates, life-size, on graph paper. You could use trial and error with a compass, or the intersecting chord theorem.

The magnetic flux density (B) at the centre of the Helmholtz coils is given by:

$$B = 9.0\,(NI/r) \times 10^{-7}\,\text{T}$$

Newton's second law of motion for the centripetal force and acceleration of the electrons gives:

$$Bev = mv^2/R$$

So

$$v = BR\,e/m$$

4 Find a value for the speed (v) of the electrons:
(i) Calculate B from the formula for the coils.
(ii) Substitute B and R in the equation giving v.
($e/m = 1.76 \times 10^{11}\,\text{C kg}^{-1}$)

Sample readings

Accelerating pd (V) = 2.40 kV
Current in the coils (I) = 0.31 A
Number of turns on each coil (N) = 320
Mean radius of the coils (r) = 7.2 cm

Co-ordinates of the deflected path

x/cm	0	2	3	4	5	6	7	8
y/cm	0	0.3	0.5	0.7	0.9	1.3	1.7	2.2

Plot the sample co-ordinates life-size to obtain the path radius, R. Calculate v.

Answers

$R \approx 15.6$ cm; $v \approx 3.4 \times 10^7\,\text{m s}^{-1}$.

Evaluation

The fluorescent line that shows the curved path of the electron beam is not sharp; the electrons have a spread of speeds and hence path radii; and the screen is only marked in centimetre squares. The measured radius of the path is therefore not precise and could have an uncertainty of 1 cm in 15 cm ($\approx$ 7%).

The value for B will be uncertain because it uses measurements of current and coil diameter. Also, it will not be exactly uniform over the whole electron path.

The overall uncertainty in the electron speed will be at least 10%.

10.4 Electric fields

Computer simulation 1: Efield
Computer simulation 2: Pfield

(There are no lab experiments for this topic.)

Computer simulation 1 Efield

Aim **To investigate the electric field strength and electric potential within a uniform electric field.**

Apparatus
- computer running the program 'Efield' from the CD

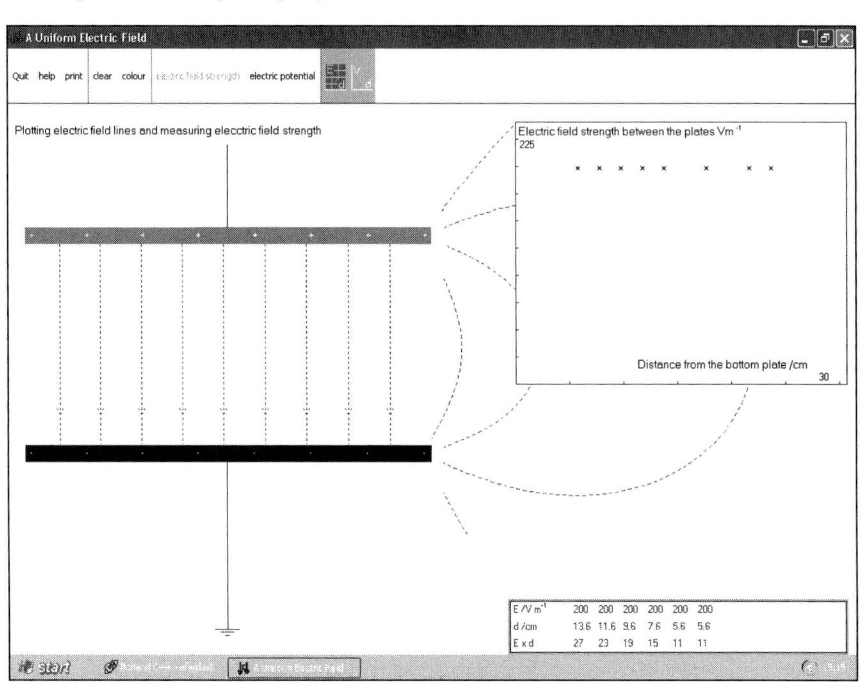

Plan
- Select 'electric field strength' and plot electric field lines by clicking the mouse in the field. Look at the graph of electric field strength E against d, and note that the field strength is constant in the region between the plates.
- Select 'electric potential' and measure the electric potential V at different positions between the plates. Look at the graph of V against d, and note that the potential rises uniformly with distance from the bottom plate.
- Print out your results and write a brief report about this topic.

Computer simulation 2 Pfield

Aim **To investigate the electric field strength and electric potential in the field of a point charge.**

Apparatus

- computer running the program 'Pfield' from the CD

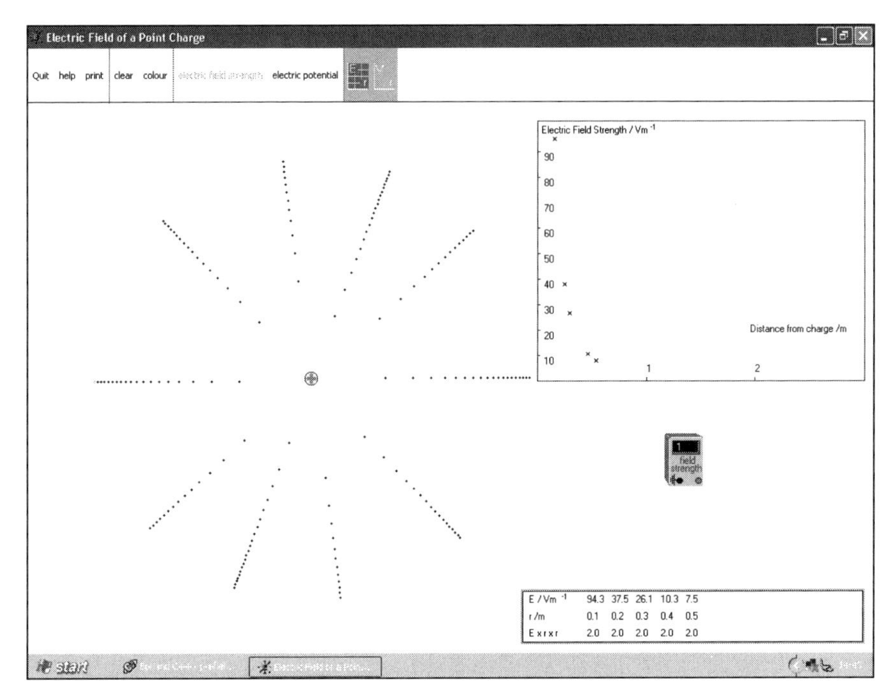

Plan

- Select 'electric field strength' and plot electric field lines by clicking the mouse in the field. Look at the graph of electric field strength E against r to see how the field strength varies with distance from the central charge. Print out your results.
- Select 'electric potential' and measure the electric potential V at different points in the field. Look at the graph of V against r, and print out your results.
- Compare how electric potential and electric field strength vary with distance from the charge.
- Write a brief report about this topic.

10.5 The deflection of moving charged particles by electric and magnetic fields

Experiment: Observing the path of electrons in 'crossed' electric and magnetic fields and measuring their speed
Computer simulation: Qpaths

Experiment: Observing the path of electrons in 'crossed' electric and magnetic fields and measuring their speed

Apparatus As for the experiment in topic 10.3 (p. 113), plus:

- second high voltage power supply, for the electric deflecting plates (see the *Note* on p. 113 about care needed with an HT supply)

If this is not available, the pd for the electric deflecting plates can be taken from the first EHT unit, as in Figure 10.9, although this will limit the control of the path.

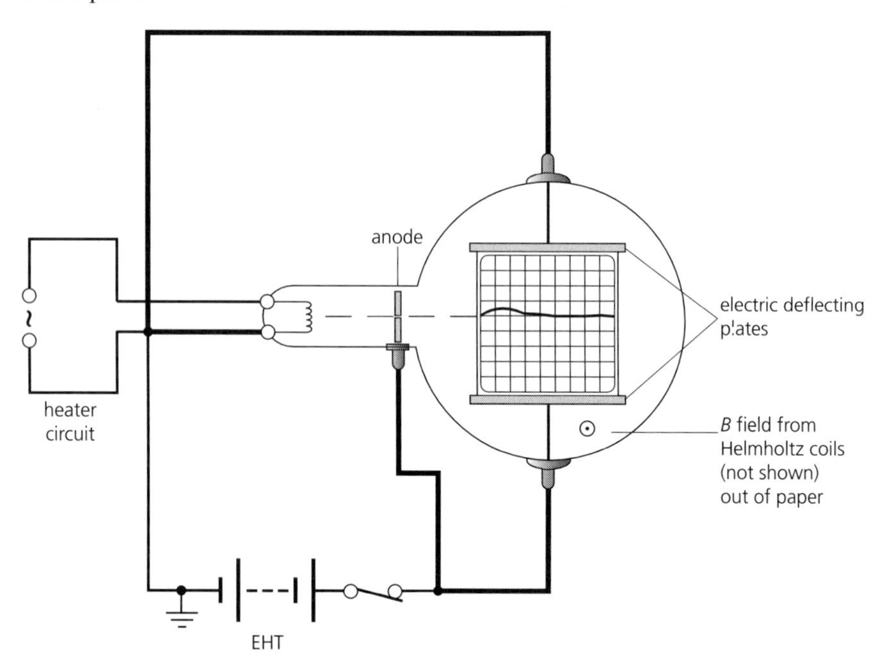

Figure 10.9

Plan

- Ensure that the high voltage supplies are switched off while connections are made.
- Check also that the supply intended for the deflecting plates is turned down to zero. Carefully connect the positive and negative terminals to the

connections of the two parallel metal deflecting plates. Turn on the power and gradually increase the pd between the plates.

- Observe the deflected path of the electrons in the electric field. Note that it is a parabola and not the arc of a circle.
- Turn on the Helmholtz coils and adjust the strength and direction of the magnetic field until the electric deflection is cancelled and the beam travels in as straight a line as you can achieve.
- Take readings that will allow you to calculate the strength of the magnetic field (see the experiment in topic 10.3, p. 114).
- Record the pd (V) and the distance between the parallel deflecting plates (d).
- Switch the supplies off before dismantling the apparatus.

Analysis

1 Record all your readings with proper descriptions and units.

Force on the electrons due to the magnetic field: $F = Bev$
Force on the electrons due to the electric field: $F = Ee$
When these forces are equal and opposite: $Ee = Bev$
and $v = E/B$

2 Find a value for the speed (v) of the electrons:
(i) Calculate the magnetic flux density: $B = 9.0\,(NI/r) \times 10^{-7}$ T
(ii) Calculate the electric field strength: $E = V/d$
(iii) Calculate a value for the speed of the electrons: $v = E/B$

Sample readings

For the electric field:
distance between plates (d) = 5.0 cm
pd between plates (V) = 2.40 kV

For the magnetic field that cancelled the electric deflection:
current in Helmholtz coils (I) = 0.31 A
mean radius of coils (r) = 7.2 cm
number of turns (N) = 320

Calculate v from these sample readings.

Answer $v \approx 4.1 \times 10^{7}\ \mathrm{m\,s^{-1}}$.

Evaluation

This is a less precise way of measuring electron speed than the experiment in topic 10.3. The lack of uniformity of the two fields makes it difficult to get the electrons to follow a straight path. The current in the coils can be varied over a wide range and the path still considered to be straight. The current could have an uncertainty of 0.05 A in 0.31 A (≈17%).

All the other measurements – pd, plate separation, radius of the coils – will add uncertainties making a total of at least 20%.

Computer simulation

Qpaths

Aim **To use electric and magnetic fields to deflect moving charged particles.**

Apparatus

- computer running the program 'Qpaths' from the CD

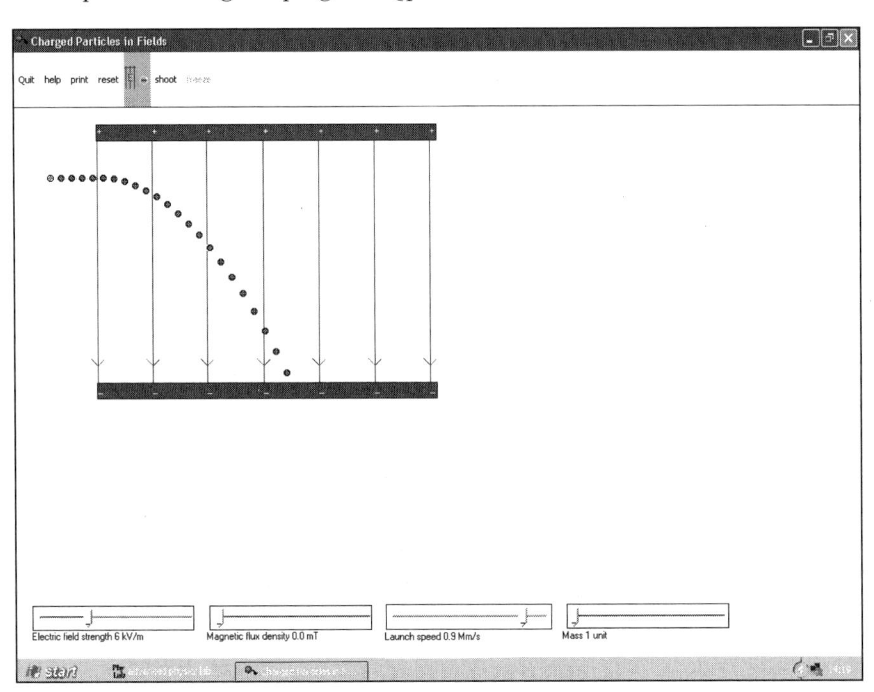

Plan

- Click on 'shoot', and the charged particle will move into the electric field and be deflected.
- There are sliders that control the strengths of the electric and magnetic fields, the speed of the particle and its mass. You can work with a positive or negative charge and you can reverse the direction of the electric field. You can change the starting position of the charge by clicking in a different place. Work systematically with these variables and explore the following.
- With an electric field only, explore:
  - (i) the effect of the speed on the path;
  - (ii) the effect of the mass on the path;
  - (ii) the effect of the direction of the field on the path.
- With a magnetic field only:
  - (i) see if you can make the particle move in a closed circular path;
  - (ii) explore the effect of speed and mass on the radius of this circle.
- With 'crossed' electric and magnetic fields:
  - (i) see if you can choose the field strengths and speed so that the particle passes through in a straight line ($E/B = v$ in this case);
  - (ii) see how complicated a path you can produce with the two fields.

Analysis Print copies of the paths of the particle in the fields and on each one put arrows to show the direction of the forces that act. Write an equation for each force and describe how speed and mass affect the shape of the path.

10.6 Electromagnetic induction

Preliminary work: Inducing an emf by (a) cutting magnetic flux and (b) changing magnetic flux linkage
Full investigation: Does the output of an alternator follow Faraday's law of electromagnetic induction?
Computer simulation: Induce

Preliminary work

Inducing an emf by (a) cutting magnetic flux and (b) changing magnetic flux linkage

Apparatus

For plan (a):

- U-shaped magnet (e.g. as found in a motor kit)
- 1 m insulated copper wire
- analogue ammeter (0–100 μA) or spot galvanometer
- dc generator/motor

For plan (b):

- U-shaped magnet
- small search coil with many turns (e.g. Maplin sealed reed relay)
- analogue voltmeter (0–100 mV)
- cycle alternator
- oscilloscope
- datalogger

Plan (a)

- Strip the insulation from the ends of the copper wire and attach it to the ammeter.
- Move a section of the wire across the magnetic field of the U-shaped magnet. You may be able to see a small movement of the ammeter needle or light spot.
- Disconnect the wire and wind it round to make a loop with several turns. Re-attach the ends to the ammeter and move a section of this loop through the field. The induced current should now be big enough to register on the meter.

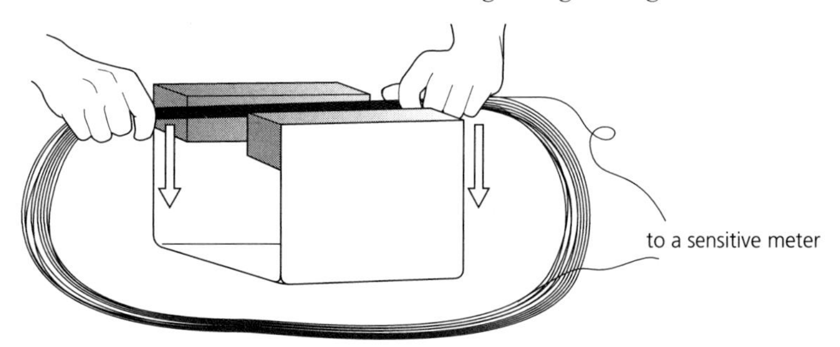

Figure 10.10

- Change the speed of your movement, then move the magnet instead of the wire, and then move the wire along the field lines, and see how the induced current is affected.

- Connect the dc generator/motor to the ammeter and slowly rotate the armature. The rotating coil inside will cut the flux of a magnet and generate a current. (You can also consider that a current is generated because the flux *linkage* through the coil is changing.)

Plan (b)

- Connect the search coil to a voltmeter.
- Change the magnetic flux linkage through the coil by moving it into (and out of) the magnetic field of the magnet. Also try rotating the coil in the field.

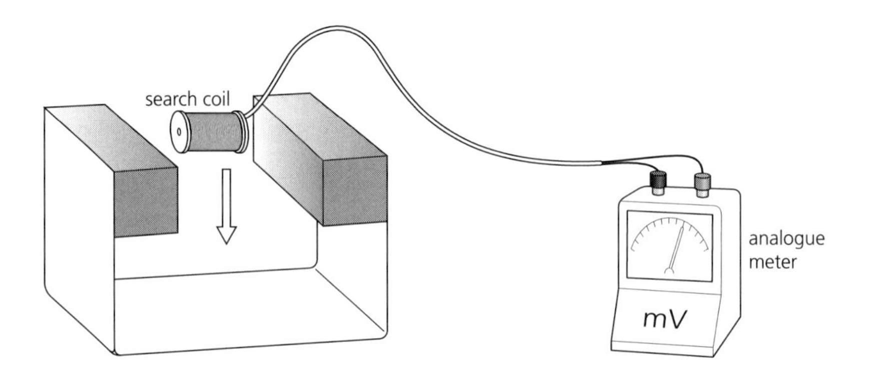

Figure 10.11

- Connect the cycle alternator to the voltmeter and turn it slowly. The rotating magnet inside changes the magnetic flux linkage through a coil and induces an emf.
- Connect an oscilloscope instead of the voltmeter. With its time base on, its display will show a graph of emf (induced by your actions) against time. The Y sensitivity should be set to about 10 mV/cm and the time base to 1 s/cm.
- If you have a datalogger, attach the search coil to its inputs, set the sampling speed to 'every 5 ms', and obtain a permanent record of the induced emf as you cut flux or change flux linkage.

Analysis

Draw diagrams of the methods you used to induce a current or emf. For each, say whether it seems magnetic flux is being cut or flux linkage changed.

Sample readings

A datalogger was used to record these readings of induced emf as a small coil was moved out of the magnetic field of a U-shaped magnet. The first extraction took 120 ms; the second 200 ms.

Very quick change

time /ms	0	5	10	15	20	25	30	35	40	45	50	55	60
emf /mV	4	4	4	4	4	4	7	85	164	179	149	118	92
time /ms	65	70	75	80	85	90	95	100	105	110	115	120	
emf /mV	72	58	50	45	39	32	27	23	19	15	13	11	

Quick change

time /ms	0	5	10	15	20	25	30	35	40	45	50	55	60
emf /mV	4	4	4	4	4	4	4	4	4	4	4	4	4
time /ms	65	70	75	80	85	90	95	100	105	110	115	120	125
emf /mV	11	21	32	41	49	56	61	68	72	75	78	78	76
time /ms	130	135	140	145	150	155	160	165	170	175	180	185	190
emf /mV	73	67	60	50	44	37	33	27	22	13	16	14	12
time /ms	195	200											
emf /mV	10	9											

Plot a graph of emf (y-axis) against time (x-axis) for both sets of sample readings, on the same axes.

The results for the 'very quick change' show a large emf for a short time, and the 'quick change' a smaller emf but for a longer time. The area under the graph gives the change in magnetic flux linkage and should be the same for both.

Estimate the area under each peak to see if the results bear this out.

Full investigation

Does the output of an alternator follow Faraday's law of electromagnetic induction?

Apparatus The equipment available determines the method used.

For plan (a):

- cycle alternator on a bicycle
- digital or analogue ac voltmeter
- frequency meter or stop watch
- connecting leads
- datalogger (if available)

For plan (b):

- alternator (or dynamo) that can be turned by hand
- large capacitor to smooth the output if a dynamo is used
- oscilloscope
- connecting leads for the oscilloscope
- datalogger (if available)

For plan (c):

- alternator that can be turned by hand
- datalogger
- connecting leads

Plan (a)

- Support the bicycle upside down on the bench. Push the alternator knob onto the tyre and disconnect the lead to the lamps.
- Connect the voltmeter and the frequency meter (if used) between the alternator output and its case.
- Use the pedals to turn the wheel steadily and check that the meters give readings.

- Turn the pedals and hold the emf at as steady a value as you can. Read the frequency meter or, if you are using a stop watch, time 20 rotations of the pedals (t). Find the number of rotations of the alternator (n) for one rotation of the pedals. The frequency of the induced emf is $20\,n/t$.
- Repeat for 5 other emfs and rotation speeds.

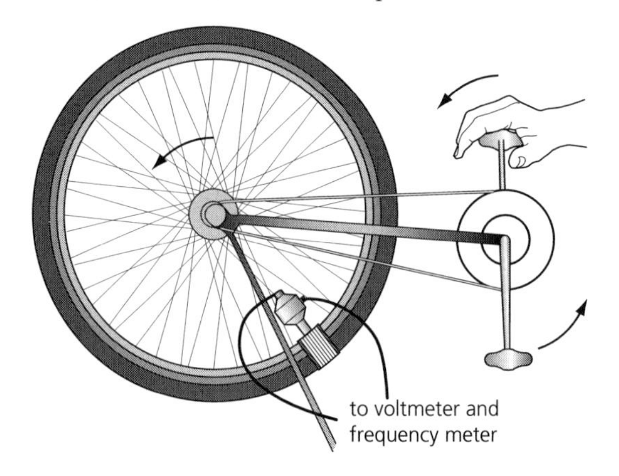

Figure 10.12

Analysis Put your results into a table and plot a graph of the induced emf against frequency.

The frequency of rotation of the alternator is proportional to the rate of change of magnetic flux through the alternator coil. If the graph of emf against frequency is a straight line through the origin, it shows that the alternator does follow Faraday's law.

Sample readings These results were taken using a laboratory dynamo that could be turned by hand. A large capacitor was connected across the dynamo to smooth its output and maintain the peak emf across the voltmeter.

induced emf /V	1.0	2.0	3.0	4.0	5.0	6.0	7.0	8.0	9.0	10.0
no. of rotations	10	20	20	30	30	30	30	40	40	40
time taken /s	45.3	44.8	29.6	33.3	26.2	23.2	19.7	22.8	20.5	18.3
frequency /Hz										

Calculate the frequency of rotation of the dynamo for each set of sample readings. Plot a graph of induced emf against dynamo frequency.
Report on whether the results show that emf and frequency are proportional.
By considering the structure of a dynamo, explain how the frequency of rotation is related to the rate of change of magnetic flux linkage through its coil.

Evaluation The chief difficulty lies in holding the output emf steady at a chosen value. The varying output introduces uncertainties in the values of emf and frequency. To estimate the overall random uncertainty, calculate emf/frequency for each set of readings and find the variation from the average. The graph of emf against frequency will show whether the quantities are proportional.

Plan (b)

- Connect the output from the alternator or dynamo to the input of the oscilloscope.
- Ask a colleague to turn the alternator at a steady speed while you adjust the Y-amplifier control and time base of the oscilloscope to give a stable trace of the output. Make sure the time base central control is set on 'calibration'.
- Hold the speed steady and measure the peak voltage of the waves and their distance apart. Try to get an average of both quantities. Note the time base setting and the Y-amplifier setting.
- Repeat at different rotation speeds. The waves should get taller and the distance apart shorter as the speed rises.

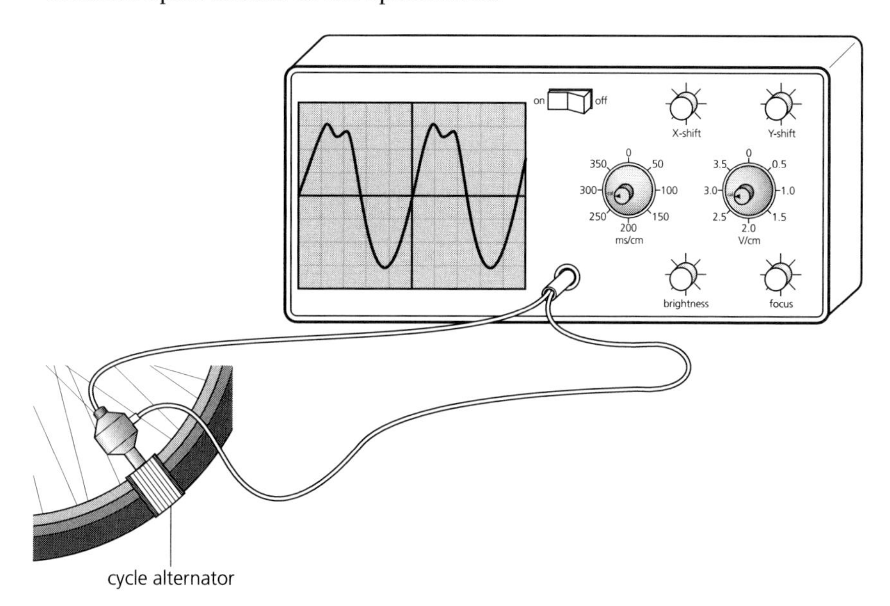

Figure 10.13

Analysis

1 Record your results in a table including the following entries:

Mean height of voltage waves /cm
Y-amplifier setting /V cm^{-1}
Peak induced emf /V
Mean distance between wave crests /cm
Time base setting /s cm^{-1}
Time for one wave /s
Frequency /Hz
Peak emf/frequency /V s

2 Plot a graph of peak induced emf against frequency. If this is a straight line through the origin, the alternator follows Faraday's law.

Evaluation To estimate the uncertainty, calculate peak emf/frequency for each set of readings and find the variation from the average.

Plan (c)

- Connect an alternator to the voltage inputs of a datalogger.
- Set the sampling rate to 5 ms and turn on the 'trigger' if there is one.
- Roll the knob of the alternator along the bench at a steady speed. Cloth or paper on the bench can help to stop the knob slipping.
- Practise a few times until you have a good trace of the induced emf. Extract average values of the peak emf and the time between peaks.
- Repeat for different rates of roll.

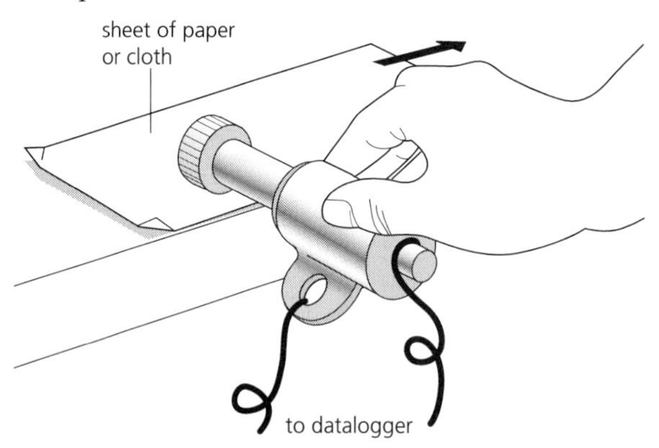

Figure 10.14

Analysis

1 Record average emf, time between voltage peaks and frequency in a table.
2 Plot emf against frequency to look for proportionality.

Sample readings

The following results were obtained by a datalogger connected to a cycle alternator. Values of emf were recorded every 5 ms as the knob of the alternator was rolled steadily along the edge of the bench. The peak voltages and the time intervals between them were then extracted from the data. An example of one of the datalogger traces is shown in Figure 10.15. (The datalogger could not measure negative voltages, so half of the peaks of the alternating emf are cut off.)

average peak emf /mV	2000	1800	1500	1000	700
average time between voltage peaks /ms	36	41	50	78	110
frequency of rotation of alternator /Hz					
emf/frequency /V s					

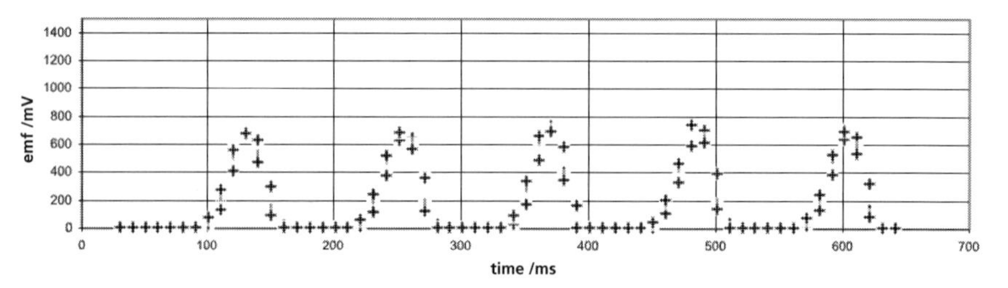

Figure 10.15

Calculate frequency and emf/frequency to check for proportionality. Plot a graph of emf against frequency. Use the graph to help you conclude whether the alternator follows Faraday's law of electromagnetic induction and whether there are any anomalous readings.

Computer simulation

Induce

Aim **To investigate the emf induced when a conductor cuts magnetic flux, or when the magnetic flux linkage through a conducting loop changes.**

Apparatus

- computer running the program 'Induce' from the CD

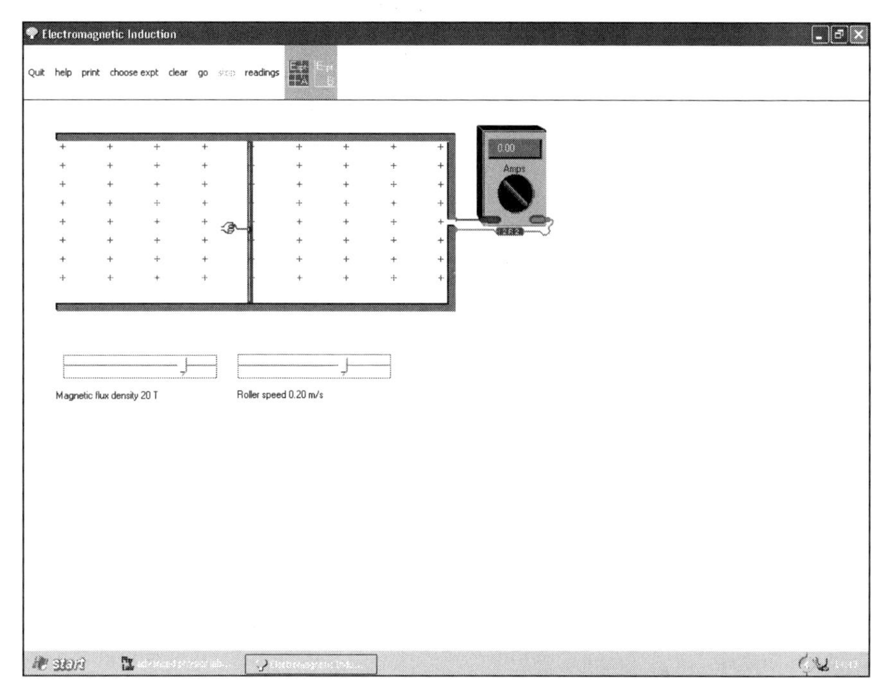

Plan

- Choose Experiment A. In this imaginary experiment a metal roller can be moved along metal rails to cut magnetic flux.
- Use the sliders to choose a magnetic flux density and a roller speed.
- Click on 'go', and an emf is generated while the roller is moving.
- Click on 'readings' to see the values, and click on the 'Expt A' graph button to see a plot.
- Change the flux density and roller speed systematically (including negative values) to see how the emf varies with each variable.
- Print a copy of your results and the graph.
- Now do Experiment B. The roller is fixed in this experiment but the flux through the loop can be made to change.
- The sliders control two variables: the size of the change in magnetic flux density and the time for that change. Together they control the 'rate of change of magnetic flux linkage'.
- Again work systematically with the variables to investigate how the induced emf depends on the rate of change of flux linkage.
- Print out your results and the graph.

11 Oscillations

11.1 Simple harmonic motion

Preliminary work: Setting up simple harmonic motion (SHM)
Computer simulation: SHM
Investigation: Observing the simple harmonic motion of a mass on a spring and finding the spring constant

Preliminary work — Setting up simple harmonic motion (SHM)

Apparatus

- 2 light springs
- 2 weight hangers and extra masses
- 2 half-metre rules
- 2 pendulum bobs
- thread
- masking tape
- small wooden or plastic blocks
- clamps, bosses and stands

Plan The idea is to build four oscillating systems and observe their motion.

- Clamp the top of a spring between two blocks and support them from a clamp stand. Hang a weight hanger on the spring and set it off making small vertical oscillations.
- Clamp and support the top of another spring and hook a weight hanger on the free end. Secure the hanger to the spring with masking tape. Set off the weight hanger making small radial oscillations about a vertical axis.

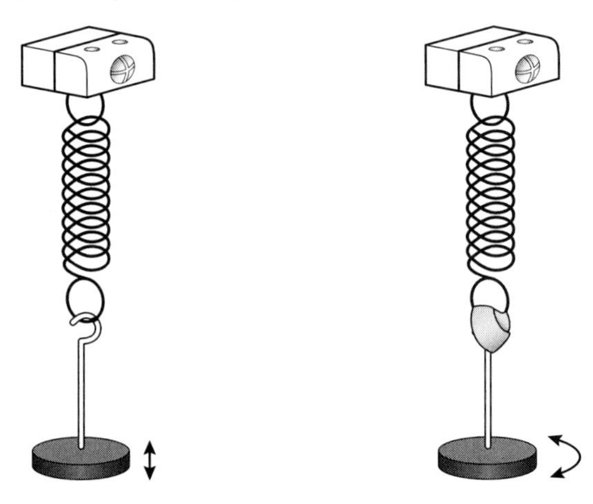

Figure 11.1

- Clamp a half-metre rule horizontally between two clamp stands and hang a second half-metre rule horizontally from the first by two parallel threads. Set the bottom rule vibrating with small horizontal radial oscillations about a vertical axis. See Figure 11.2.

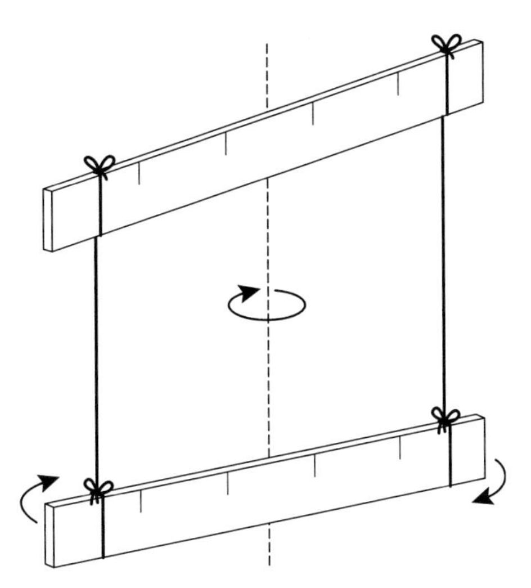

Figure 11.2

- Make two pendulums with bobs and thread. Clamp the top of the thread of each one between two blocks fixed to stands. Place the pendulums side by side and join their threads by another thread (see Figure 11.3). Set one swinging with small sideways oscillations.

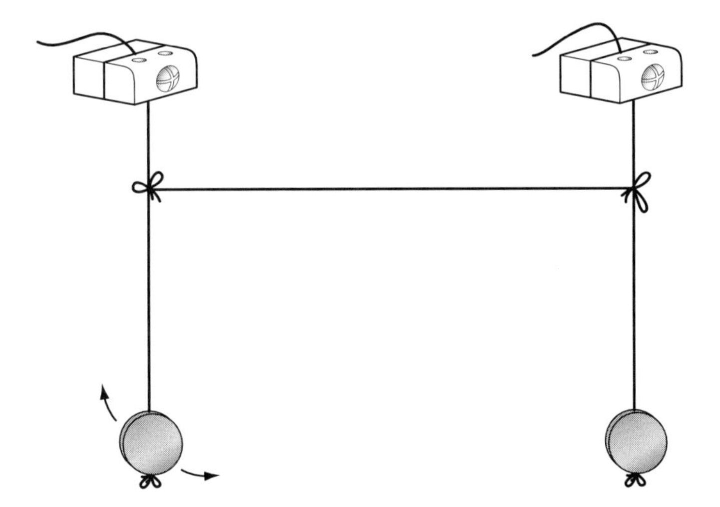

Figure 11.3

Analysis For each vibrating system, describe the motion and predict the variables that affect its period of oscillation.

Extension Do a detailed investigation of the motion of one of the systems.

Computer simulation

SHM

Aim **To investigate the effect of restoring force and mass on the acceleration, velocity and period of a simple harmonic motion.**

Apparatus

- computer running the program 'SHM' from the CD

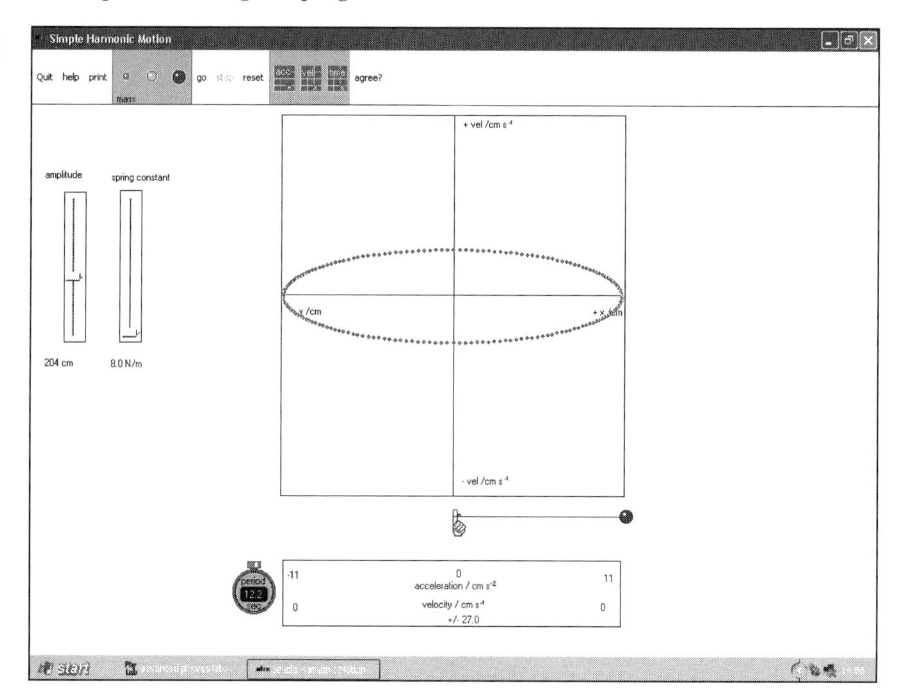

Plan The screen shows a bead on a piece of elastic held by a finger. The bead can be made to move horizontally with simple harmonic motion (SHM), miraculously unaffected by gravity.

- Click on 'go' and watch the motion. Look at the graphs of speed, acceleration and displacement, and link them to the bead's motion.
- Change the mass of the bead, the 'spring constant' of the elastic and the amplitude of motion to see if they change the motion in the way you would predict.
- Click on 'agree?' and tick the boxes you agree with. Print out a copy ('My conclusions') for your notes.

Investigation

Observing the simple harmonic motion of a mass on a spring and finding the spring constant

Apparatus

- light spring
- weight hanger and 6 masses (100 g)
- stop clock or watch
- plastic or wooden blocks
- clamp, boss and stand

Plan

- Clamp the top of the spring in two plastic blocks supported by a clamp stand, as in the Preliminary work (p. 128).

- Hang the weight hanger and masses on the spring and set them into small vertical oscillations. Let the oscillations settle.
- Arrange a viewing mark behind the hanger, so that you can accurately judge the centre of an oscillation. Count '0–0–0–0' each time the hanger passes the mark in the same direction.
- On one of the '0' counts, start the stop clock and continue counting '1–2–3–' up to 40 oscillations. Stop the clock on '40' and record the reading. (For rapid oscillations increase the number to 60 to make a total time greater than 30 s.)
- Do one check reading of the time for this mass.
- Change the mass, and time oscillations as before.

Skill level (Implementing)

A: I secured the mass and spring to give small, one-dimensional oscillations. I located a viewing mark at the centre of the oscillation and timed and counted oscillations. I used a larger number of oscillations when the period was short. I did one check reading for each mass. I found the average period for 6 different masses.

All but one of the above = B; all but two = C; all but three = D; all but four = E.

Analysis

1 Draw a table for your results, with extra columns for processing. An example is shown in the *Sample readings* below. You could use a spreadsheet to do the processing.

Theory predicts that:

$T^2 = 4\pi^2 m/k$ (T = period, m = mass, k = spring constant)

So, for any particular spring, T^2/m should be constant. A graph of T^2 against m should be a straight line through the origin.

2 Calculate T^2/m and plot the graph of your readings to check the prediction above. Note that your best-fit line may not pass through the origin, due to a small systematic uncertainty – see the *Evaluation* overleaf.

3 Calculate a value for the spring constant:

$k = 4\pi^2 m/T^2$ (for the individual readings)

$k = 4\pi^2/\text{gradient}$ (for the graphical method)

Sample readings

mass, m /kg	no. osc.	time /s	check time /s	av period, T /s	T^2 /s^2	T^2/m /s^2 kg^{-1}
0.1	60	23.17	23.08	0.385	0.149	1.49
0.2	60	32.23	32.36	0.538	0.290	1.45
0.3	60	39.31	39.56	0.657	0.432	1.44
0.4	40	30.17	30.36	0.757	0.572	1.43
0.5	40	33.40	33.67	0.838	0.703	1.41
0.6	40	36.52	36.83	0.917	0.841	1.40
0.7	40	39.26	39.27	0.982	0.964	1.38
					std	0.03

Plot the graph suggested in the *Analysis* for these sample readings. Find the gradient of the best-fit line. Do not assume it goes through the origin. Calculate a value for the spring constant.

Answer ~29.5 N m^{-1}.

Evaluation In the sample readings the times are recorded to 4 figures, as shown by the stop watch used, but are only significant to 3 figures because of human reaction time. The calculated quantities are therefore given to 3 significant figures. The random uncertainties can be estimated from the differences from the average, or the standard deviation, and are about 2%.

There is a systematic error that has been ignored. Part of the spring is moving and about a third of its mass should have been included in the mass value. So the overall error would be greater than 2%.

Improving the plan

- Use additional mass values in between 0.1 kg and 0.2 kg, and 0.2 kg and 0.3 kg, because the period changes rapidly for the smaller masses.
- Time more oscillations – the calculated periods will then be more precise.
- Allow for the moving mass of the spring. When $T^2 = 0$, the intercept on the mass axis gives the effective mass of the spring that should be added to each load mass in the calculations of T^2/m.

Extension To make this a full investigation, after finding the spring constant in this way, use a static method (such as that in the experiment of topic 7.2, p. 59) for the same spring. Compare and evaluate the results.

11.2 Damped oscillations

Experiment: Measuring the change in amplitude of oscillations that are subject to frictional forces
Computer simulation 1: Damped oscillations (1)
Computer simulation 2: Damped oscillations (2)

Experiment Measuring the change in amplitude of oscillations that are subject to frictional forces

Apparatus

- light spring
- bar magnet
- weight hanger and masses
- large diameter solenoid with many turns
- ammeter (100 μA) or spot galvanometer
- connecting leads
- clamp, boss and stand
- beaker to fit inside the solenoid
- water and cooking oil
- datalogger (if available)

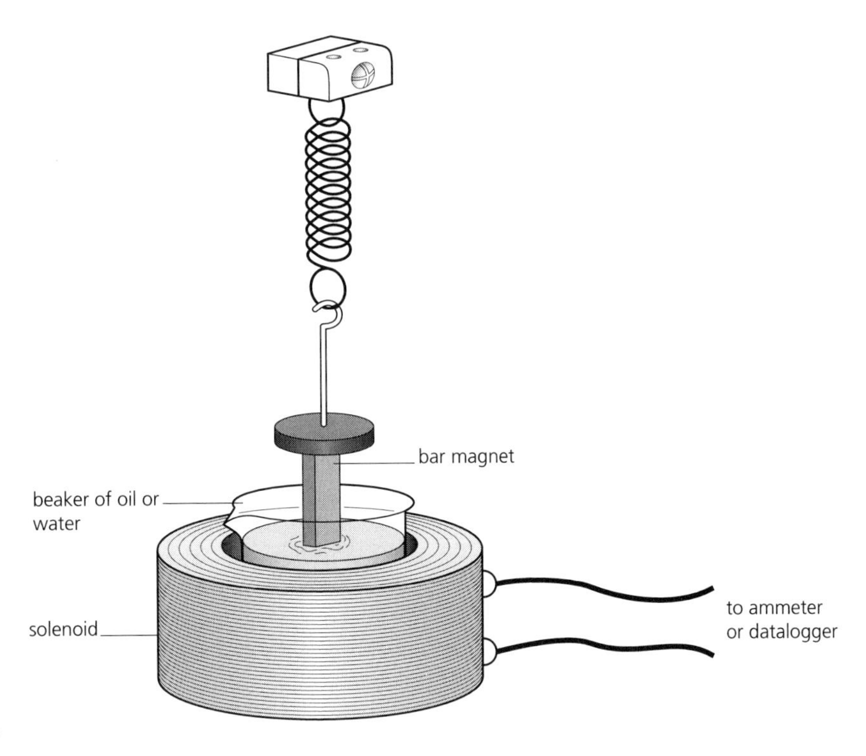

Figure 11.4

Plan

- Attach the magnet lengthways underneath the weight hanger.
- Hook the hanger onto the spring and position them so that the magnet oscillates inside the mouth of the solenoid.
- Connect the ammeter directly to the solenoid to register the current induced by the movement of the magnet.
- Set the magnet oscillating. Check that the current follows the oscillations and that the maximum value of the current decreases as the oscillations die away.
- Take readings of the maximum current for every oscillation if you can.
- Put a small beaker of water inside the solenoid so that part of the magnet oscillates in the water. The water will cause more friction than the air and so damp the motion.
- Obtain maximum current readings for every oscillation of this motion.
- Replace the water with oil to increase damping still further, and obtain readings to show the way the amplitude of the motion dies away.

Alternative plan using a datalogger

It is difficult to read the fast-moving deflections of the ammeter by eye. A datalogger can give more frequent and more precise readings of the induced emf.

- Connect the solenoid to the inputs of the datalogger and record the voltage variations as the oscillations die away. A suitable setting for a bar magnet on a single spring is a reading every 50 ms for 30 s.
- Choose the option that shows a graph of pd against time on the screen, and print out a copy. Alternatively, to plot a graph of the decaying amplitude by hand, extract peak pd and time values from the data. This has been done in the *Sample readings* overleaf.

Analysis

1 Tabulate your readings neatly and plot graphs of maximum current against time for the three degrees of damping. (The maximum induced emf and maximum current occur at the centre of the oscillations where the velocity is greatest. For SHM the maximum velocity is proportional to the amplitude, and so the maximum current is proportional to the amplitude of the motion.)
2 Suggest why the loss in amplitude decreases as the amplitude gets smaller.

Sample readings

Peak induced emfs, recorded by a datalogger, of a bar magnet on a spring bouncing in oil, water and air

time /s	0	0.5	1.0	1.5	2.0	2.5	3.0	3.5	4.0	4.5	5.0	5.5	6.0	6.5
peak emf in oil /mV	100	77	64	45	41	31	27	20	17	15	9	9	6	5
peak emf in water /mV	100	92	85	80	73	70	63	60	54	49	43	41	40	36
peak emf in air /mV	100	96	87	85	81	77	77	72	68	68	64	63	60	60

time /s	7.0	7.5	8.0	8.5	9.0	9.5	10.0	10.5	11.0	11.5	12.0	12.5	13.0	13.5
peak emf in oil /mV	3	3												
peak emf in water /mV	33	32	27	27	25	22	22	18	22	18	13	13	12	11
peak emf in air /mV	58	51	54	54	51	45	45	45	45	42	40	38	37	37

time /s	14.0	14.5	15.0	15.5	16.0	16.5	17.0
peak emf in oil /mV							
peak emf in water /mV							
peak emf in air /mV	40	37	33	32	32	32	32

Plot the amplitude figures given in the sample readings for the three levels of damping, on the same axes.

Computer simulation 1 Damped oscillations (1)

Aim **To plot damped oscillations for different degrees of damping using a spreadsheet, such as Excel.**

Apparatus

- computer running Excel or similar program

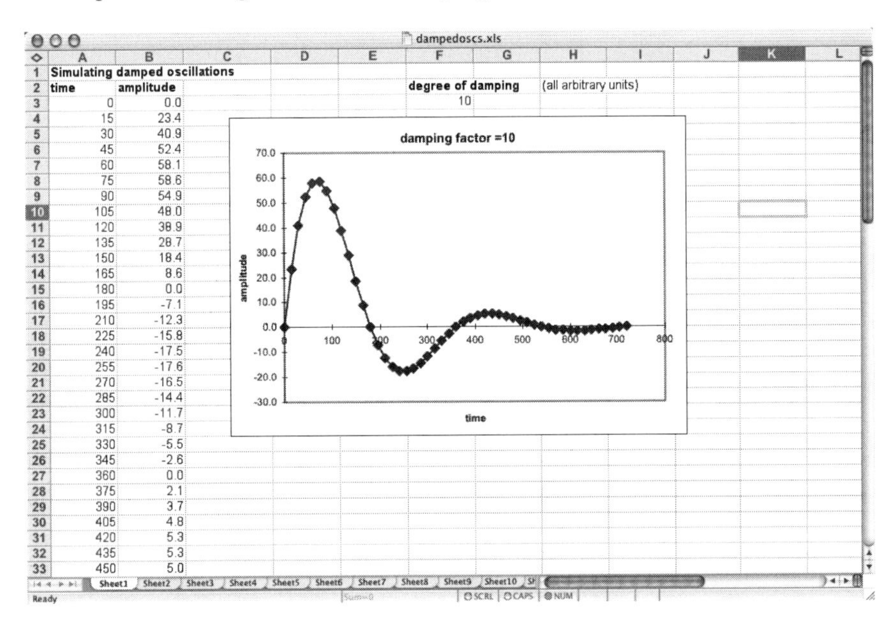

Simulating damped oscillations

time	amplitude
0	0.0
15	23.4
30	40.9
45	52.4
60	58.1
75	58.6
90	54.9
105	48.0
120	38.9
135	28.7
150	18.4
165	8.6
180	0.0
195	-7.1
210	-12.3
225	-15.8
240	-17.5
255	-17.6
270	-16.5
285	-14.4
300	-11.7
315	-8.7
330	-5.5
345	-2.6
360	0.0
375	2.1
390	3.7
405	4.8
420	5.3
435	5.3
450	5.0

degree of damping (all arbitrary units)
10

Plan/Analysis

- In column A put in times 0, 15, 30, … every 15 ms up to 540 or more. You can use a formula to do this. Put 0 in A1. In A2 put **=A1+15** and 'fill down' as far as you like. (Go to Edit–Fill–Down).
- In cell B1 put this formula:

  =100*SIN(PI()*A3/180)/EXP(F3*A3/1500)

  and 'fill down' to the bottom of the times in column A.
- In cell F3 put the degree of damping: 1 = light, 10 = critical.
- Select the numbers in column B and use the Chart Wizard to plot a scatter graph.
- Change the degree of damping. The graph will show the effect immediately. Print out copies for your notes.

Skill level (Analysing)

A: I entered the time values in a headed column using the formula. I entered the formula to calculate amplitude. I filled the formula down to the bottom of the time values. I used the program to plot and print a graph of amplitude against time. I changed the degree of damping and printed off two more graphs.

All but one of the above = B; all but two = C; all but three = D; all but four = E.

Computer simulation 2 Damped oscillations (2)

Aim **To plot amplitude/time graphs of damped oscillations for different degrees of damping.**

Apparatus

- computer running the program 'Pullit' from the CD

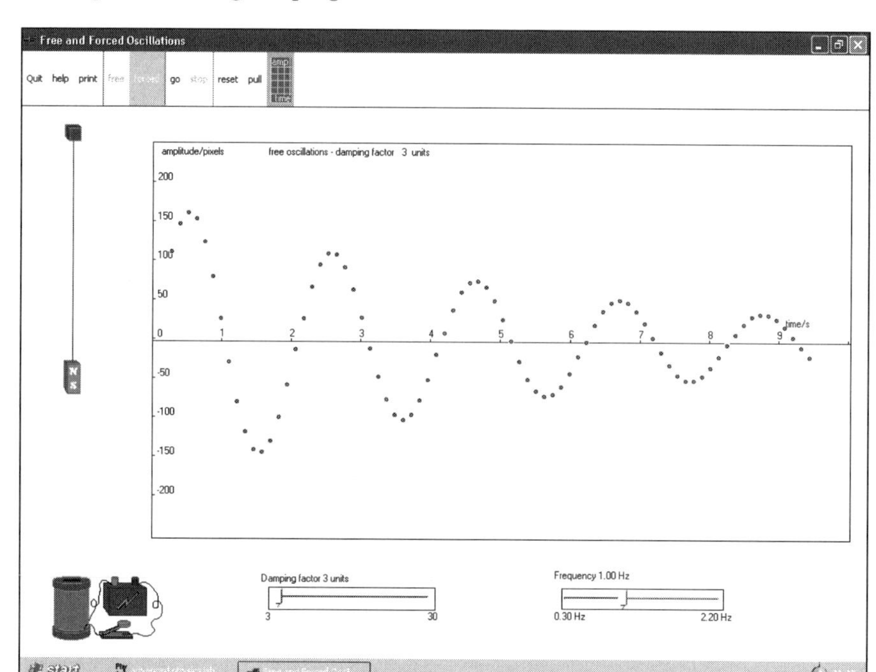

Plan

- Select 'free' and click on 'go'. The magnet will start to bounce on its piece of elastic. Watch the amplitude die away slowly as energy is lost to the air.
- Change the degree of damping using the 'damping factor' slider and start again.
- Print a copy of the plot of damped oscillations for at least three different degrees of damping (e.g. 3, 15 and 30 units). The screen above shows the plot for light damping (3 units).

Analysis Give an example of where in everyday life you would meet each level of damping shown by your graphs.

11.3 Forced oscillations and resonance

Preliminary work: Forcing a magnet on a spring to oscillate and finding its resonant frequency
Computer simulation: Forced oscillations

Preliminary work

Forcing a magnet on a spring to oscillate and finding its resonant frequency

Apparatus

- bar magnet
- 2 or 3 light springs connected in series
- adhesive tape
- large diameter solenoid with many turns
- dc low voltage power supply and connecting leads
- press switch
- clamp, boss and stand
- stop clock or watch

Plan

- Tape the magnet long-ways to the free end of the springs and hang the top spring from a clamp stand so that the magnet can bounce vertically.
- Connect the solenoid to the power supply and switch, and place it under the magnet.
- Press the switch on and off in a regular rhythm and check that small regular forces act down-wards on the magnet.
- Change your rhythm to try to match the natural frequency of the magnet and build up large amplitude oscillations.
- Hold this resonant frequency while a partner times 20 oscillations.
- Change over, and time your partner's attempts at achieving resonance.

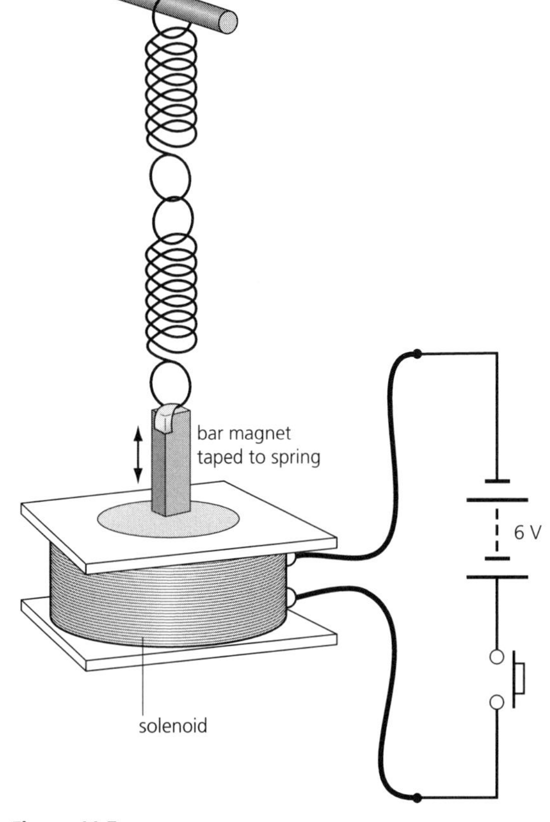

Figure 11.5

Analysis

1 Write a paragraph explaining how a small periodic force can make a system vibrate; how normally the amplitude of vibration is small but, if the force has the same frequency as the natural frequency of the vibrating system, large amplitude vibrations can build up.
2 State your average value of the resonant frequency of your system.

Computer simulation

Forced oscillations

Aim **To produce forced oscillations at different frequencies and to plot a resonance curve.**

Apparatus

- computer running the program 'Pullit' from the CD

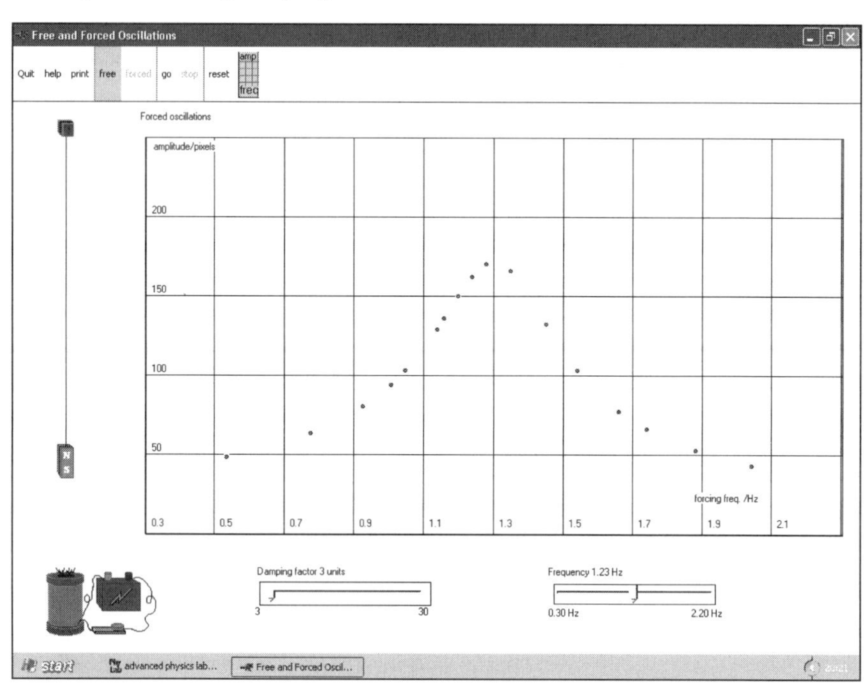

Plan

- Choose 'forced', click 'go', and the switch turns the electromagnet on and off at regular intervals. The magnet starts to move with forced oscillations and a point is plotted on the amplitude/frequency graph.
- Move the 'frequency' slider to alter the switch frequency.
- Work through the whole range of frequency to build up a graph of amplitude against forcing frequency.
- Move the 'damping factor' slider to change the damping, and repeat.
- Print a copy of the plot of forced oscillations for at least three different levels of damping.

Analysis

1 Write a paragraph to go with your graphs, explaining the meaning of the shape of the curves and how damping affects the sharpness of resonance.
2 Explain also, at which points in the cycle the switch should be pressed on and then off to make the magnet resonate.

12 Waves

12.1 Waves on a spring

Preliminary work 1: Transferring energy by a transverse wave
Preliminary work 2: Transferring energy by a longitudinal wave
Preliminary work 3: Making two transverse waves combine to form a stationary wave
Computer simulation: Twave

Preliminary work 1 Transferring energy by a transverse wave

Apparatus

- long spring (e.g. 'slinky') with a piece of brightly coloured wool tied to one of its coils

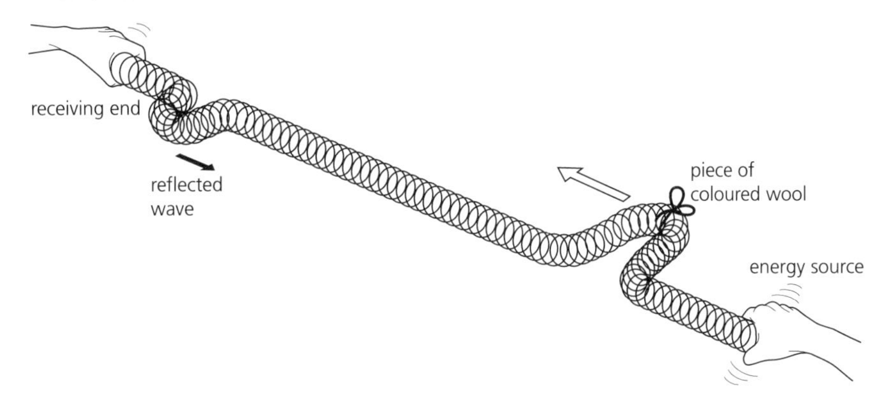

Figure 12.1

Plan

- Work with a partner.
- Lay the spring on a flat surface with one person (the energy source) holding one end and the other person (the receiver) holding the other.
- Ask the person supplying the energy to make a 'hump' (displacement) in the spring by flicking his or her wrist up and down or from side to side.
- Watch the piece of wool as the displacement passes by.
- Watch the receiver's hand to see if it vibrates when the displacement arrives.
- Watch the displacement as some of the wave energy is reflected from the receiving end.
- Make smaller and larger 'humps' to see if the energy put in (or amplitude) changes the speed of the wave.
- Change the tension in the spring to see if that changes the speed of the wave.

Analysis Write a report, describing:

- the movement of the displacement ('hump') and the movement of the coils of the spring as the displacement passes by;
- the most effective way you found of transmitting energy to the receiver;
- what happens to the displacement when it is reflected;
- how to change the speed of the wave along the spring.

Preliminary work 2 Transferring energy by a longitudinal wave

Apparatus As for Preliminary work 1.

Plan

- Ask the person supplying the energy to send a compression along the spring by moving his or her hand inwards (along the spring) and then outwards.
- Make the same observations as before.

Analysis Report on your observations, explaining what you see as the compression pulse passes along. Describe the motion of the piece of wool. Explain how you can make the longitudinal pulse move faster or slower.

Preliminary work 3 Making two transverse waves combine to form a stationary wave

Apparatus As previously.

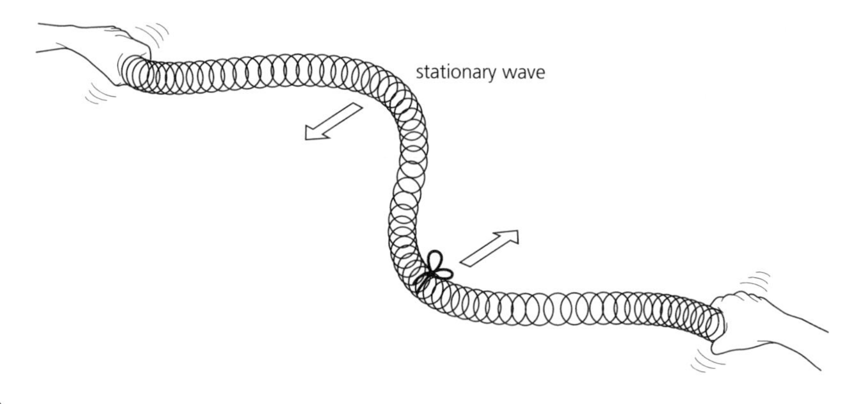

Figure 12.2

Plan

- The people at the ends of the spring should both send continuous transverse waves towards each other. They must use the same frequency as each other and adjust that frequency until the loops of a stationary wave are seen. There should be crests that do not move along (*antinodes*) and places where there is no movement (*nodes*).
- Make the spring vibrate in its *fundamental* mode (one loop) and in the *first* and *second harmonic* modes (two loops and three loops).

Analysis Make careful drawings of the stationary waves that you see. Mark the positions of the nodes and antinodes.

Computer simulation

Twave

Aim **To understand how a wave travels through a medium.**

Apparatus ■ computer running the program 'Twave' from the CD

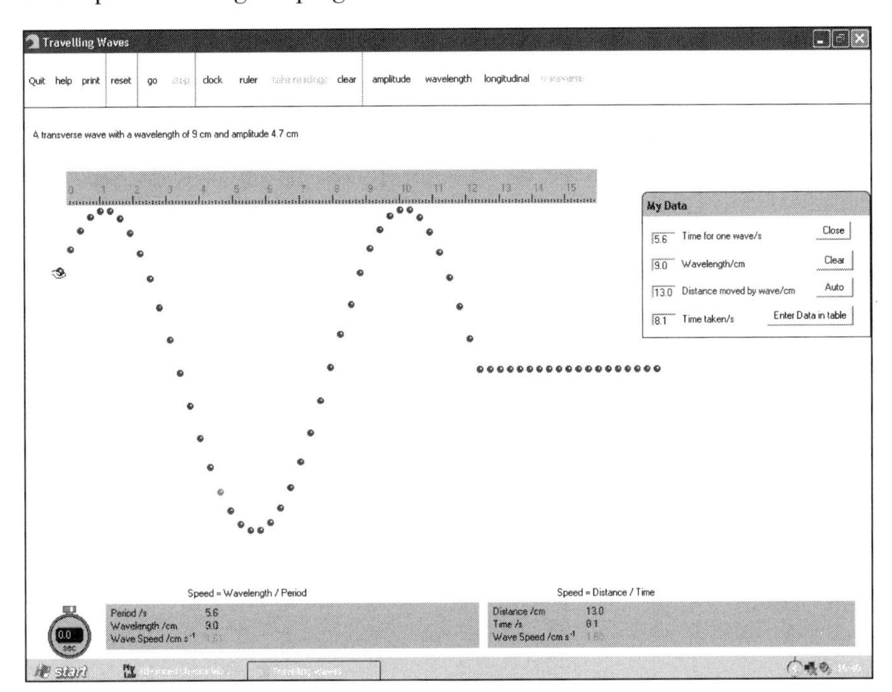

Plan

- Click 'go' and watch the beads as the wave builds up. Notice how each bead (watch the red one) moves with vertical simple harmonic motion. All the beads eventually reach the same amplitude and frequency but have a *different phase* from their neighbours. A bead reaches its maximum displacement slightly later than the one to its left. The result of this is to produce a wave crest that moves along the string of beads.
- Make waves of different amplitudes and wavelengths.
- Stop the wave and use the ruler and clock to take readings. Enter readings into the data table to calculate the speed of the wave. Either:

  speed = wavelength/period or:
  speed = distance/time

  can be used. They should give the same result!
- Change to 'longitudinal' and watch how a longitudinal wave progresses along the string of beads.

Analysis

1. Print out pictures of the waves and any readings that you have taken.
2. Explain the difficult idea of how a wave transmits energy through a material without wholesale movement of that material.

12.2 Sound waves

Preliminary work: Measuring the speed of sound using echoes
Experiment: Measuring the frequency of a sound wave using an oscilloscope
Investigation: Measuring the wavelength and speed of sound in free air

Preliminary work — Measuring the speed of sound using echoes

Apparatus

- long measuring tape
- stop watches
- hammer and block of wood
- large open outdoor space with a reflecting wall
- thermometer

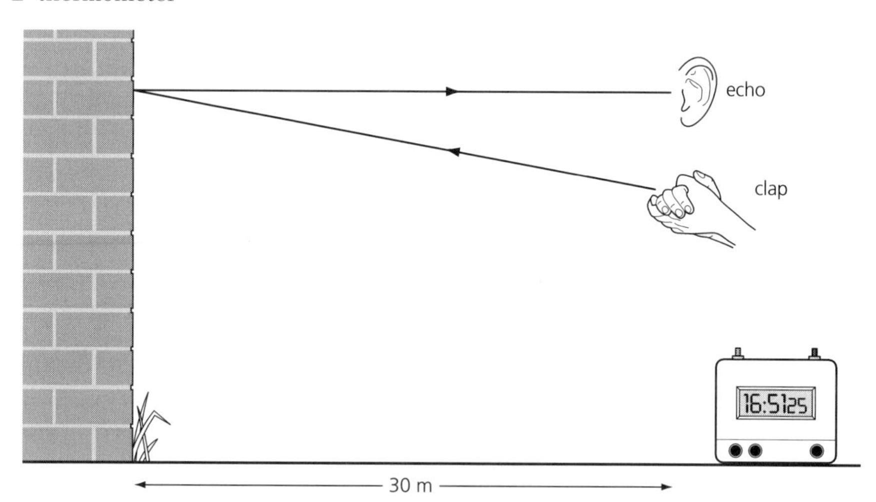

Figure 12.3

Plan

- Work in a group and find a wall that gives a distinct echo.
- Stand about 30 metres from the wall and measure the distance carefully with the measuring tape.
- Make a sharp sound by clapping, or hitting the block of wood with the hammer. Listen for the echo. The sound travels to the wall and back in less than a second – too short a time to measure with a stop watch.
- As soon as you hear the echo, clap/bang again to make a new sound. Adjust the timing of the claps/bangs until the echo of one is 'covered' by the next new sound, and seems to disappear. (Another method is to alternate *clap–echo–clap–echo–clap* in a steady rhythm. The sound will then have travelled 4 times to the wall and back between claps.)
- Keep the rhythm going while other members of the group time 64 claps: count down from 8 to zero, start the watches, count 8 lots of 8 then stop the watches. (There is a natural musicality about working in 8s.) Record all of the watch readings and repeat a number of times to get a good average.
- Record the air temperature.

Analysis

1 Write down distance and time readings. Find the average of the times, leaving out any that appear anomalous.

Total distance travelled in the time = 64 × 2 × distance to wall (or 64 × 4 × distance to wall, depending on the method used)

2 Calculate the speed = distance/time. Quote it to a suitable number of significant figures, together with the air temperature.

Sample readings

Air temperature = 25 °C
Distance to wall = 31.7 m
Average times for 64 claps using the 'disappearing echo' method:

run 1: 13.2 s run 2: 13.7 s run 3: 12.1 s

Use these sample readings to work out a value for the speed of sound in air at 25 °C.

Answer 312 m s^{-1}.

Evaluation

This 'order of magnitude measurement' is an example of an experiment that is accurate but not precise. The distance measurement can be made with precision (an uncertainty of only about 10 cm in 30 metres ≈ 0.3%) but the time measurement cannot. This depends on finding and keeping to the correct rhythm and then timing it by hand. The uncertainty could be 1 s in 12 s (≈ 8%). It is also easy to miscount, but this would be a blunder and give an anomalous reading.

Improving the plan

- Use a datalogger or other electronic method for registering the echoes but still retain the technique of regularly repeating sounds to extend the total journey distance.

Experiment: Measuring the frequency of a sound wave using an oscilloscope

Apparatus

- oscilloscope
- audio-signal generator
- loudspeaker
- microphone
- connecting leads

Plan

- Connect the loudspeaker to the signal generator, switch on and select a frequency that gives a pleasant note.
- Connect the microphone to the input of the oscilloscope. Adjust the control knobs of the oscilloscope until you get a stable trace of the sound wave on the screen. (To find the trace, it is often helpful to press all the buttons out – except the 'on' button – and to connect the input switch to 'ground'.)

- Choose a time base setting that gives at least one complete wave on the screen. Make sure its 'fine' control is set to 'calibration'.
- Use the X-shift knob to position the wave carefully, and use the centimetre grid to measure from one peak to the next. Note down this distance and the time base setting.
- Click to higher time base settings and get readings of the distance between as many peaks as possible.

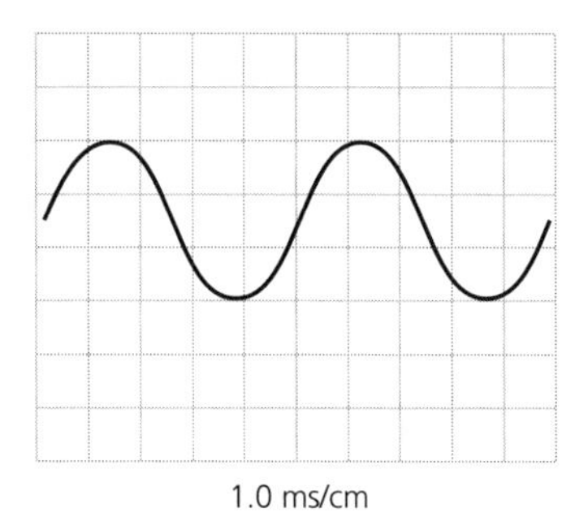

Figure 12.4

Analysis

1 Put the readings into a table. An example is given in the *Sample readings* below.

Time it takes to 'draw' the waves on the screen
= time base setting × distance the waves cover on the screen.

2 Calculate the time it takes to 'draw' one wave (the period) and from this calculate the frequency (1/period).
3 Work out an average value for the frequency of the sound.

Sample readings

number of waves	distance on grid /cm	time base setting in ms /cm	period /ms	frequency /Hz
1	5.6	0.1	0.560	1786
3	8.4	0.2	0.560	1786
7	7.9	0.5	*	*
15	8.4	1.0	*	*
			average	*

Calculate the missing values (*) of period and frequency.
Calculate the average frequency and round it down to 2 significant figures as in the measured data.

Answer 1800 Hz.

Evaluation The trace on the oscilloscope is about 0.5 mm thick and the grid is marked in 0.2 cm units. The grid lies in front of the screen and so parallax can add uncertainty to the reading. Together these give an uncertainty in the distance measurement of about 0.2 cm in 8 cm (≈ 2%). The time base calibrations are given to 1 decimal place which suggests quite a large uncertainty (although it may have been difficult to print too many numbers on the control panel). The intermediate calculations have been made to 3 significant figures and then rounded down to 2 as determined by the readings.

Investigation: Measuring the wavelength and speed of sound in free air

Apparatus

- double-beam oscilloscope
- audio-signal generator
- loudspeaker
- microphone
- connecting leads
- metre rule
- masking tape
- clamp and stand
- thermometer

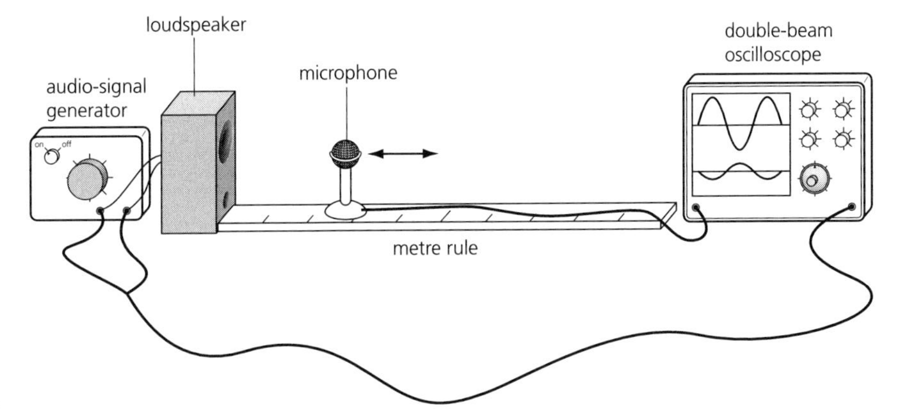

Figure 12.5

Plan

- Connect the microphone to one input of the oscilloscope and the audio-signal generator to the other.
- Connect the audio-signal generator to the loudspeaker.
- Tape a metre rule to the bench at right angles to the loudspeaker.
- Clamp the microphone to a stand, placed so that it can slide along the metre rule.
- Use the signal generator and loudspeaker to make a sound of about 5000 Hz and position the microphone to receive it.
- Adjust the oscilloscope controls until you have two waves on the screen – one from the microphone and the other from the signal generator.
- Check that as you move the microphone away from the loudspeaker, one trace moves relative to the other on the screen.
- Start with the microphone close to the loudspeaker and adjust its position until the traces on the screen line up. Record this position on the metre rule.
- Move the microphone away from the loudspeaker until the waves on the screen line up again. Use the wave traces to judge when you have moved the microphone exactly one wavelength through the sound wave. Continue to move the microphone through as many wavelengths as you can.
- Record the number of wavelengths moved through and record the microphone's new position.
- Measure the frequency of the waves as in the previous experiment.
- Record the air temperature.

Skill level (Implementing)

A: I was able to connect the apparatus together and get two wave traces on the oscilloscope without help. I chose an audio frequency that gave a strong signal from the microphone. I moved the microphone until the two wave traces were aligned. I moved the microphone an exact number of wavelengths and measured the distance moved. I measured the frequency of the sound wave.

All but one of the above = B; all but two = C; all but three = D; all but four = E.

Analysis

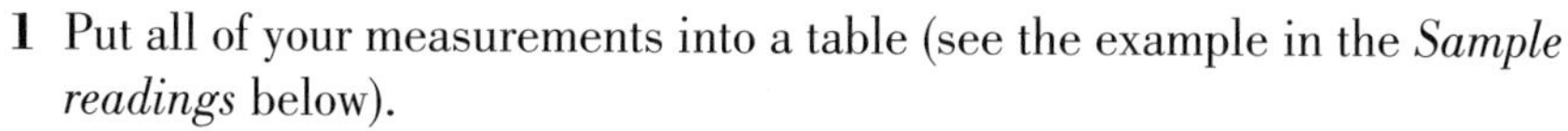

1 Put all of your measurements into a table (see the example in the *Sample readings* below).
2 Calculate a value for the wavelength of the sound.

Wave speed = frequency × wavelength

3 Calculate a value for the speed of sound in air.
4 Explain carefully why the length of the waves on the oscilloscope screen is not the wavelength of the sound: the trace represents a pressure/time graph of the wave as it passes the microphone.
5 Explain also why one trace moves relative to the other when you move the microphone through the sound wave in the air.

Sample readings Air temperature = 24 °C

first microphone position /mm	second microphone position /mm	distance moved /mm	number of wavelengths	wavelength /mm	frequency /Hz	speed /m s^{-1}
869	447	422	6	70.3	5200	366

Evaluation This is a direct but not very precise way of measuring the speed of a travelling sound wave in free air. It is difficult to judge the two microphone positions exactly and there could be an uncertainty of 1 cm in each. When these are subtracted they give a smaller quantity with increased % uncertainty (2 cm in 70 cm or ~3%). Adding an uncertainty of 2% from the determination of the frequency (see p. 144) gives an overall uncertainty of ~5%.

Improving the plan

- Find the frequency that gives the largest signal from the microphone. Cheap microphones do not have flat responses and resonate at certain frequencies. A large signal from the microphone makes it possible to move through more wavelengths before the trace gets too small, and so reduces the uncertainty in this quantity.
- Use a frequency meter to measure the sound frequency.

Extension To make this a full investigation, measure the wavelength and speed of the sound wave for different frequencies.

12.3 Stationary waves

Preliminary work: Finding resonant frequencies that make stationary waves on a cotton thread
Computer simulation: Swave
Experiment: Making stationary electromagnetic waves and measuring their frequency
Full investigation: Measuring the speed of sound by setting up longitudinal stationary waves in a column of air

Preliminary work

Finding resonant frequencies that make stationary waves on a cotton thread

Apparatus

- vibration generator
- audio-signal generator
- connecting leads
- length of black thread
- clamp stand and rubber band
- stroboscope (if available)

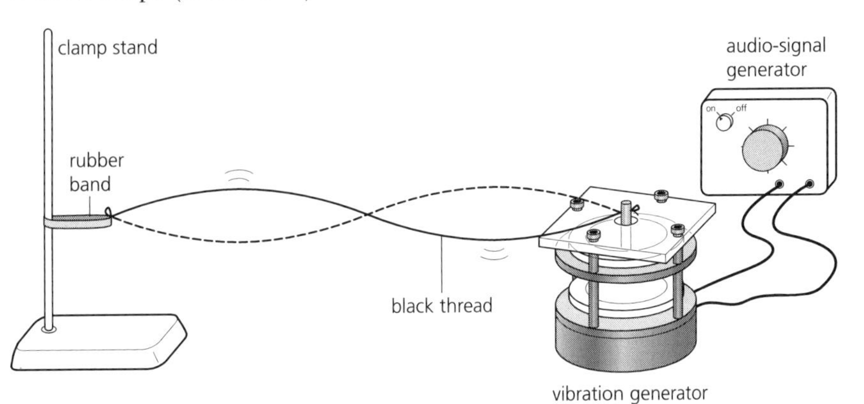

Figure 12.6

Plan

- Use the audio-signal generator to activate the vibration generator.
- Tie one end of the thread to the vibration generator and the other to the rubber band that has been hooked over the clamp stand (see Figure 12.6).
- Tension the thread by moving the stand and vibrator apart.
- Change the frequency of the generator until you see a single-loop stationary wave on the thread (the fundamental mode). Tune the frequency carefully to get maximum amplitude.
- Increase the frequency and find other modes of vibration. You should be able to get the second and third harmonic vibrations.
- Use the flashing light from the stroboscope to seemingly slow down the motion of the thread, so that you can see how it moves in detail.

Analysis

Draw careful pictures of the first three modes of vibration of the thread, showing the nodes, antinodes and wavelength of the waves. Use arrows to indicate the direction of motion of the different sections of the thread.

Computer simulation

Swave

Aim **To watch the build-up of a stationary wave from two waves travelling in opposite directions.**

Apparatus

- computer running the program 'Swave' from the CD

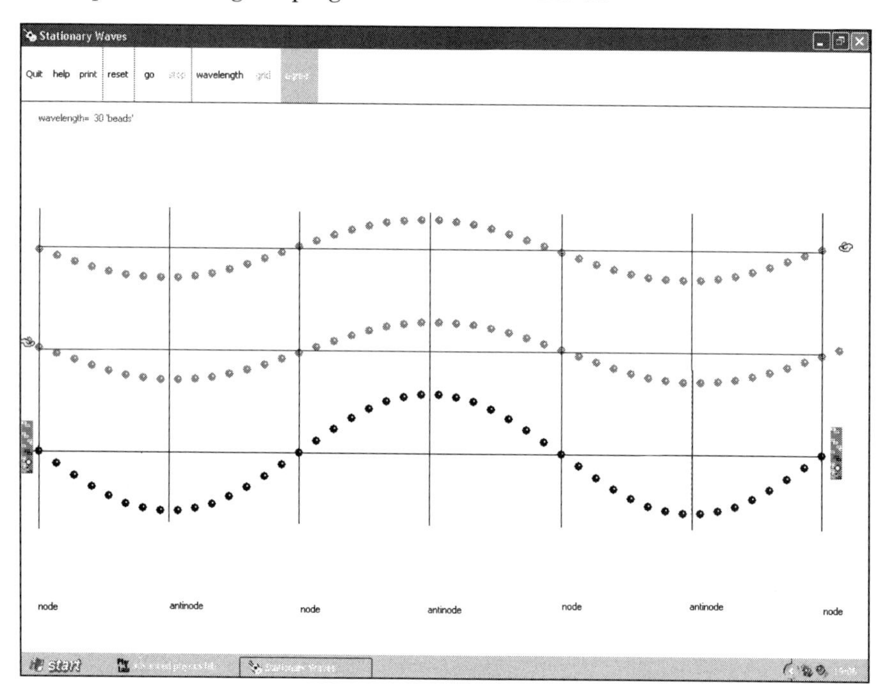

Plan

- Click 'go' and watch how a travelling wave is reflected and combines with itself to form a stationary wave, seen building up on the green beads.
- Switch on the 'grid' and, when the stationary wave has developed, 'stop' the motion in three positions.
- Change the wavelength and repeat.
- Fill in the 'agree?' dialogue box.
- Click 'print' to obtain three drawings of the frozen stationary wave and the statements you agreed with.

Analysis

1 Write a report on the motion of the green beads, referring to the following facts:

- all the beads move in phase but have differing amplitudes;
- the stationary waves have crests that do not move along the string of beads;
- the waves store but do not transmit energy;
- at the nodes the two travelling waves have equal and opposite displacements at all times, and so interfere destructively to give zero resultant displacement;
- at the antinodes the displacements are always equal and in the same direction, and so the two waves interfere constructively.

2 Include your printout(s), which will give some helpful sentences for your report.

Experiment Making stationary electromagnetic waves and measuring their frequency

Apparatus

- 3 cm microwave transmitter (note that old ones with Klystron valves use dangerous voltages and should be discarded or re-built)
- probe detector
- 100 μA ammeter
- connecting leads
- metal sheet supported vertically by wooden feet or in a clamp stand
- metre rule
- masking tape

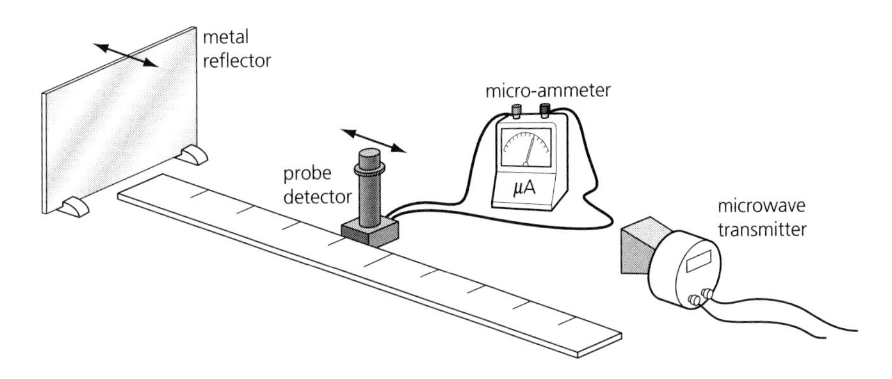

Figure 12.7

Plan Anyone with an artificial pacemaker should stand well away from the microwave transmitter.

- Switch on the transmitter, connect the probe to the micro-ammeter and check that a signal is received. Some transmitters have to be tuned to optimise the signal.
- Tape the metre rule to the bench, place the transmitter at one end and the metal sheet (reflector) at right angles to the other end (see Figure 12.7).
- Move the probe along the rule and find a 'node' where the signal is a minimum.
- Adjust the transmitter–reflector distance slightly until the reading at the node is as small as possible.
- Note the position of the probe and then move through as many nodes as you can, counting as you go. Record the finishing position of the probe.
- Do a second run as a check.

Analysis

1. Tabulate the start and finish positions of the probe, and the number of nodes it has passed.
2. Use the readings to calculate an average value of the distance between nodes. The wavelength will be twice this distance.
3. Use the speed of electromagnetic waves in a vacuum (3.0×10^8 m s^{-1}) to calculate the frequency of the microwaves.

Sample readings

start position of probe /cm	49.8	54.4
finish position of probe /cm	82.3	83.8
number of nodes passed	20	18

Calculate an average value for the wavelength of the microwaves, using both sets of readings.
Calculate the frequency of the microwaves, assuming they travel at the speed of light, 3.0×10^8 m s^{-1}.

Answers 3.26 cm; 9.20×10^9 Hz.

Full investigation

Measuring the speed of sound by setting up longitudinal stationary waves in a column of air

Apparatus

- large measuring cylinder (at least 35 cm tall)
- 5 tuning forks of different frequencies (middle C and above)
- half-metre rule
- rubber bands
- thermometer
- wide glass or plastic tube of length about 50 cm that will go inside the measuring cylinder

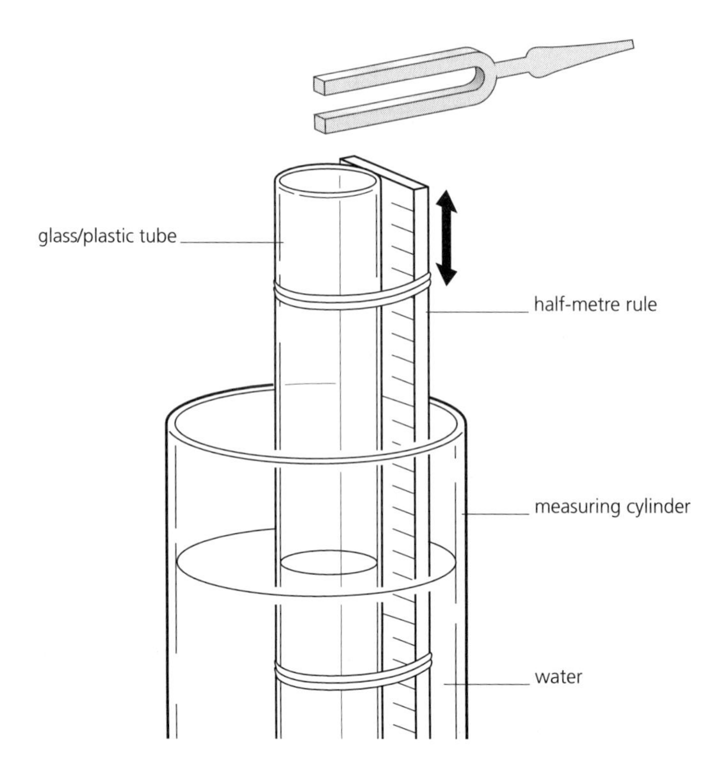

Figure 12.8

Plan

- Fill the measuring cylinder almost to the top with water.
- Use the rubber bands to fix the half-metre rule to the glass tube, with its zero mark exactly at the top end.
- Place the tube and rule in the water in the measuring cylinder.
- Strike the middle C tuning fork on your heel (or on a rubber bung) and hold it horizontally above the mouth of the tube.
- Listen to the note, and raise and lower the tube until the air column resonates – the volume of sound from the vibrating air column reaches a maximum.
- Take the reading of the metre rule at the water level. This is the length of the air column (ℓ). Note the frequency of the fork (f).
- Choose the tuning fork with the next highest frequency, find its resonant length and measure as before. Repeat with all of the tuning forks.
- Measure the temperature of the air in the tube.

Skill level (Implementing)

A: I constructed the apparatus with the end of the tube at ear level. I could strike the tuning fork correctly and make it sound. I was able to find a resonance position. I asked a colleague to help me measure the length of the air column. I found resonant lengths for 4 other forks.

All but one of the above = B; all but two = C; all but three = D; all but four = E.

Analysis

1 Record the readings of length and frequency in a table or a spreadsheet, adding two rows for the calculations of wavelength and speed (see the *Sample readings* below).

Resonance occurs when a stationary wave is formed in the air column. The first resonance occurs when the length of the air column (ℓ) is one quarter of the wavelength of the sound wave (λ). A pressure node forms at the open end of the tube and a pressure antinode at the water surface where the sound is reflected.

$\lambda = 4\ell$

2 Calculate the speed of sound ($v = f\lambda$) for each frequency.

3 Calculate an average value for the speed of sound at the temperature of the air in your experiment.

Sample readings

Air temperature = 24 °C

frequency /Hz	256	288	320	384	427	480		
length of air column /cm	33.0	28.9	25.9	21.7	19.8	17.3		
wavelength /cm	132.0	115.6	103.6	86.8	*	*		
speed of sound /m s^{-1}	338	333	332	333	*	*	average	*
							std	2.7

Calculate the missing values of wavelength and speed (*).
Calculate the average value of the speed of the sound wave.

Answer 334 ± 3 m s^{-1} at 24 °C.

Evaluation

The chief uncertainty is in the resonant length of the air column. The tube has to be positioned by judging when the sound from the vibrating air is a maximum, and then the length measured by the metre rule. There could easily be an uncertainty of 3 mm in positioning the tube and an extra 1 mm in reading the water level on the metre rule.

The overall random uncertainty can be estimated from the numerical difference of each value from the average, or the standard deviation. This is about $3\ \text{m s}^{-1}$ ($\sim 1\%$) for the sample readings.

In fact the pressure node of the stationary wave does not occur exactly at the end of the tube but a small distance outside. This introduces an unknown systematic uncertainty – or 'end error' – which makes all the readings, and the final result for speed, inaccurate (too small).

Improving the plan

- Eliminate the end error by measuring the distance from the first resonance position ($\frac{1}{4}$ of a wavelength) to the second ($\frac{3}{4}$ of a wavelength). This gives a value for $\frac{1}{2} \times$ wavelength. You will need long tubes to do this.
- Use a microphone and oscilloscope or a sound level meter to determine when the sound level is maximum.

12.4 The diffraction of waves

Preliminary work: Observing what happens to straight water waves when they pass through a narrow gap
Experiment 1: Measuring the intensity pattern of microwaves diffracted through a narrow gap
Computer simulation: Diffraction pattern
Experiment 2: Observing the diffraction of light when it passes through a narrow slit

Preliminary work

Observing what happens to straight water waves when they pass through a narrow gap

Apparatus

- ripple tank
- light source to give shadows of the waves on the ceiling or the bench
- large beaker of water
- 2 strips of wood or metal to make the gap
- long straight 'dipper' that can be vibrated at different frequencies
- hand stroboscope

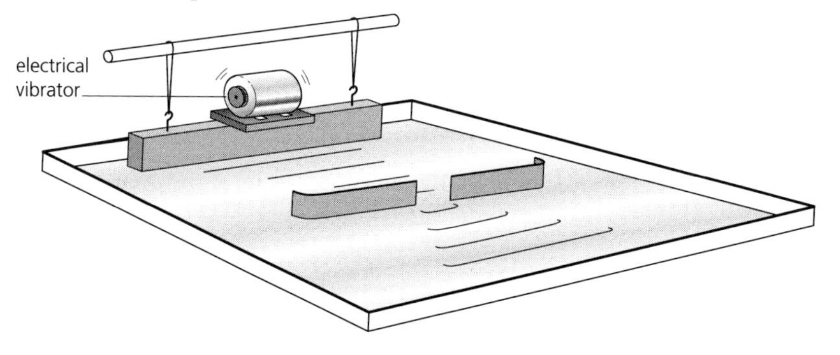

Figure 12.9

Plan

- Set up the ripple tank and pour in water to a depth of about 1 cm. Check that the waves on the water surface are visible by touching it with your finger.
- Adjust the long vibrating bar until it just touches the water surface and makes clear straight waves in the water. You can freeze the waves by looking at them through a hand stroboscope or by flicking your open fingers in front of your eyes.
- Make a narrow gap in the path of the waves from the two pieces of wood or metal. Look carefully at the waves after they have passed through the gap.
- Look for (three) differences that occur when the width of the gap is changed.

Analysis

Write an illustrated report on your findings, or copy and complete this table:

Gap width	Observation	Agree/disagree/couldn't see
more than 5 wavelengths	The waves came through straight in the middle but curved a little at the edges	
about 5 wavelengths	The wavefronts emerging were almost circular arcs, the amplitude was smaller and the pattern spread beyond the edges of the gap.	
1 or 2 wavelengths	The waves were very weak and were almost complete semi-circles spreading round almost to the barriers.	

Experiment 1 Measuring the intensity pattern of microwaves diffracted through a narrow gap

Apparatus

- microwave transmitter (see note about old-style transmitters on p. 149)
- probe detector
- ammeter (100 μA) or digital multimeter on 200 mV range
- connecting leads
- 2 metal sheets on feet or stands
- metre rule
- masking tape

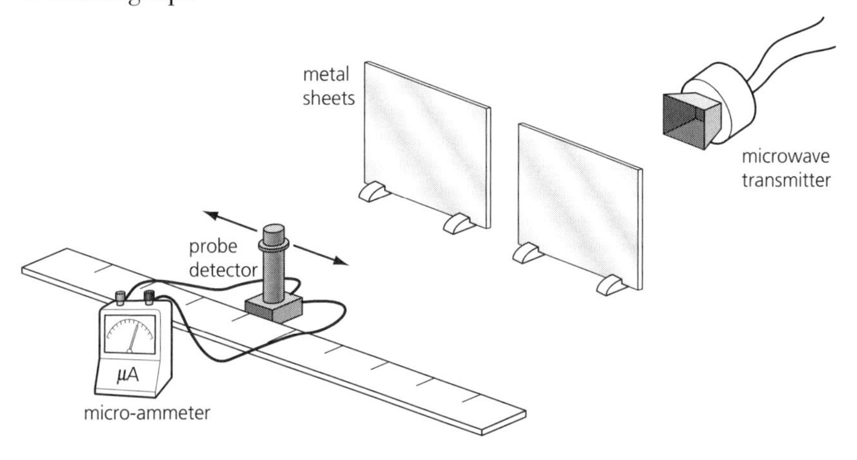

Figure 12.10

Plan Anyone with an artificial pacemaker should stand well away from the transmitter.

- Stand the two metal sheets side by side with a gap of about 12 cm between them (see Figure 12.10).
- Tape the metre rule to the bench parallel to the metal sheets and about 30 cm from them.
- Place the transmitter about 50 cm from the gap on the other side.
- Connect the ammeter to the probe, put it in line with the gap and make sure that microwave energy is received.
- Move the probe along the metre rule and take readings, every 2 cm, of the probe's position and the current or voltage it generates.
- Change the width of the gap and, if possible, get a set of readings for a narrower and a wider gap.

Analysis

1 Put the readings for each gap width into a table.
2 Plot current or voltage against distance for each gap width, on the same graph.
3 Draw as many conclusions as you can about the diffraction pattern and how the width of the gap changes the pattern.

Sample readings Gap width = 12 cm (about 4 wavelengths)
Gap to probe distance = 35.0 cm
Transmitter to gap distance = 50.0 cm

distance from centre /cm (+ to the right, − to the left)	−30	−28	−26	−24	−22	−20	−18	−16	−14	−12	−10	−8	−6	−4	−2	0
meter reading /mV	2	8	9	6	6	14	17	50	62	21	10	28	45	92	142	185

distance from centre /cm (+ to the right, − to the left)	30	28	26	24	22	20	18	16	14	12	10	8	6	4	2
meter reading /mV	1	1	2	2	8	4	4	16	39	48	38	26	58	98	148

Plot a graph of meter reading (which represents the intensity of the radiation) against the distance from the centre, for these sample readings. On the x-axis shade in a rectangle that shows the width of the gap.

Computer simulation

Diffraction pattern

Aim **To plot the intensity distribution of a diffraction pattern using a spreadsheet, such as Excel, and see the effect of gap width.**

Apparatus

- computer running Excel or similar program

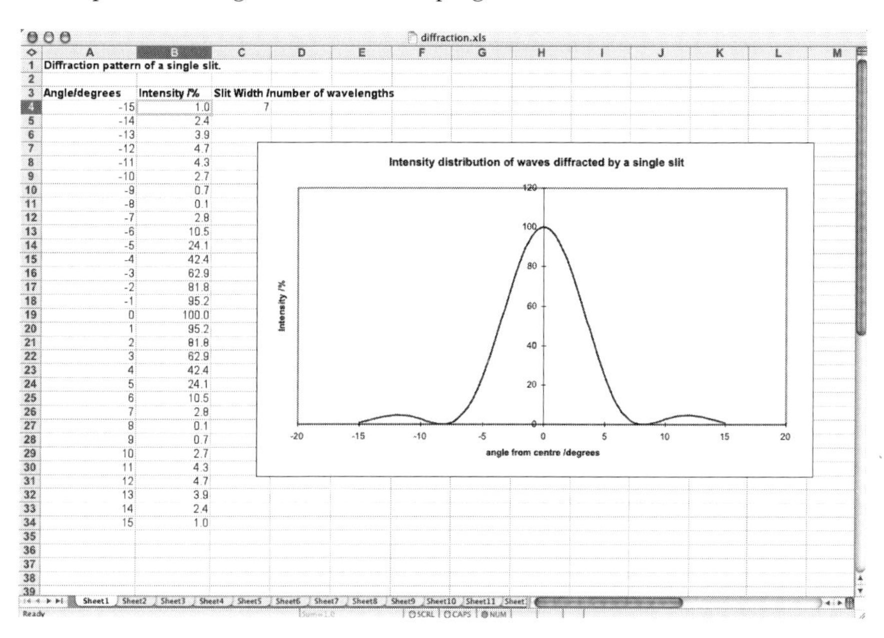

Angle/degrees	Intensity /%	Slit Width /number of wavelengths
-15	1.0	7
-14	2.4	
-13	3.9	
-12	4.7	
-11	4.3	
-10	2.7	
-9	0.7	
-8	0.1	
-7	2.8	
-6	10.5	
-5	24.1	
-4	42.4	
-3	62.9	
-2	81.8	
-1	95.2	
0	100.0	
1	95.2	
2	81.8	
3	62.9	
4	42.4	
5	24.1	
6	10.5	
7	2.8	
8	0.1	
9	0.7	
10	2.7	
11	4.3	
12	4.7	
13	3.9	
14	2.4	
15	1.0	

Plan/Analysis

- In column A, starting at A4, put in angles from −15 to +15 degrees.
- In cell C4 put the width of the gap, in wavelengths.
- In cell B4 put this formula to calculate intensity:

=100*POWER(SIN(PI()*C4*SIN(A4*PI()/180))/ (PI()*C4*SIN(A4*PI()/180)),2)

and 'fill down' from B4 to B34. (Go to Edit–Fill–Down.)

- Select the numbers in the A and B columns and use the Chart Wizard to plot a scatter graph.
- Change the gap width in C4 to see the effect on the diffraction pattern. Print out copies for your notes.

(These are theoretical graphs based on an equation derived from wave theory.)

Skill level (Analysing)

A: I entered the angle values in a headed column. I entered the formula to calculate intensity. I filled the formula down to the bottom of the angle values. I used the program to plot and print a graph of intensity against angle. I changed the gap width and printed off two more graphs.

All but one of the above = B; all but two = C; all but three = D; all but four = E.

Experiment 2 Observing the diffraction of light when it passes through a narrow slit

Apparatus

- laser (Class 2 or 3a)
- photographic slide painted black, with a narrow slit scratched in the coating, or a specially designed adjustable slit
- white screen
- LDR light sensor and datalogger (if available)

See *Topics in Safety* 3rd Edition, ASE, 2001 (p. 118) for a full discussion of the safe use of lasers in a school laboratory.

Plan

- Set the laser about 2 m from the screen but do not turn it on.
- Place the slide in a clip in front of the laser opening.
- **Instruct observers to look only at the screen** and switch on the laser.
- Make fine adjustments to the position of the slit so that the laser beam passes through and produces a diffraction pattern on the screen.
- Change the width of the slit if possible and observe the effect on the diffraction pattern.
- If a datalogger with a small light sensor is available, move it steadily across the diffraction pattern and record the variation in its light intensity.

Analysis

1. Make a drawing of the light pattern, if possible for two different slit widths.
2. If you have light intensity data, plot a graph of intensity against distance across the pattern.

Sample readings

These readings were taken by steadily moving the small light sensor of a datalogger across the diffraction pattern of a helium–neon laser.
The total distance moved was 20 cm and the datalogger time readings were converted to distances measured from the centre of the pattern.

position /cm	−10.5	−10.0	−9.5	−9.0	−8.5	−8.0	−7.5	−7.0	−6.5	−6.0	−5.5
relative intensity	22.9	23.0	24.5	25.0	24.3	23.0	24.3	26.1	26.9	26.1	24.8

position /cm	−5.0	−4.5	−4.0	−3.5	−3.0	−2.5	−2.0	−1.5	−1.0	−0.5	0.0
relative intensity	30.9	35.7	38.1	36.9	32.0	33.7	46.8	60.2	69.8	74.2	76.1

position /cm	0.5	1.0	1.5	2.0	2.5	3.0	3.5	4.0	4.5	5.0	5.5
relative intensity	75.6	70.5	60.5	47.7	30.4	29.3	34.3	35.7	31.9	25.1	27.6

position /cm	6.0	6.5	7.0	7.5	8.0	8.5	9.0	9.5
relative intensity	24.2	23.4	23.6	22.6	24.0	22.4	23.0	22.3

Plot the graph suggested in the *Analysis* for the sample readings. Underneath draw a coloured picture of the diffraction pattern, making the strength of the colouring match the intensity shown by the graph.

12.5 Two-source interference of waves

Experiment 1: Making an interference pattern with microwaves and using it to measure their wavelength
Experiment 2: Making an interference pattern with light waves and using it to measure their wavelength

Experiment 1 Making an interference pattern with microwaves and using it to measure their wavelength

Apparatus

- 3 cm microwave transmitter (see note about old-style transmitters on p. 149)
- probe receiver
- ammeter (100 μA) or digital multimeter on 200 mV range
- connecting leads
- metal sheets (2 wide, 1 narrow) on feet or stands
- metre rule
- masking tape

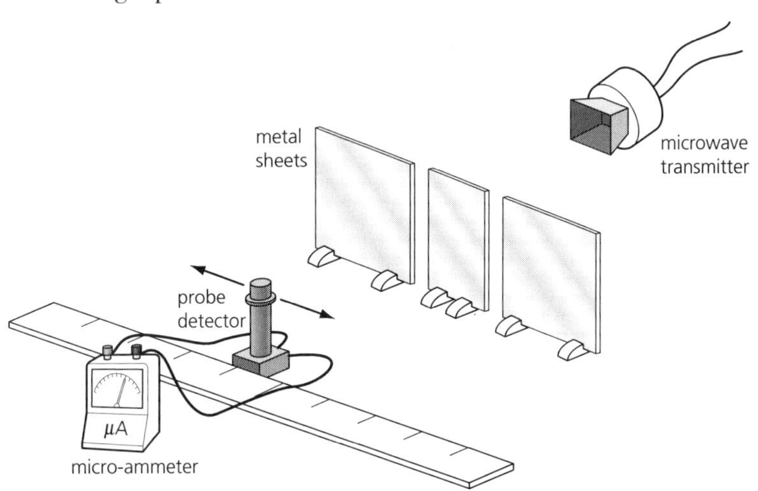

Figure 12.11

Plan Anyone with an artificial pacemaker should stand well away from the transmitter.

- Make a 'double slit' by placing the narrow sheet of metal between the other two, leaving a gap of about 2 cm between each sheet (see Figure 12.11).
- Fix the metre rule to the bench with tape, parallel to the metal plates and about 40 cm away.
- Put the transmitter about 50 cm from one side of the 'double slit', and the probe alongside the rule on the other. Connect the meter to the probe and make sure that it registers microwave energy.
- Do a test run to check that an interference pattern has been formed. The ammeter should show a series of maxima and minima as the probe is moved along the rule.

- Start again and take readings of the current or voltage every 2 cm as the probe moves across the pattern.
- Measure the exact distance from the double slit to the probe (D).
- Measure the distance between the centres of the two slits (d). (Measure between the outside edges, then between the inside edges, and take the average.)

Analysis

1. Record the readings in a table and plot a graph of probe current or voltage (representing microwave intensity) against distance across the pattern.
2. From the graph find the average separation of the maxima (Δx).

Theory shows that:

$\lambda = \Delta x\, d/D$

3. Calculate a value for the wavelength of the waves.
4. Conclude with a description of the interference pattern formed by the waves that pass through a double slit.

Sample readings Gap separation, $d = 8$ cm
Double slit to probe, $D = 35$ cm

distance from centre /cm (+ to the right, − to the left)	−30	−28	−26	−24	−22	−20	−18	−16	−14	−12	−10	−8	−6	−4	−2	0
meter reading /mV	10	1	2	2	1	2	5	23	41	23	9	7	2	9	33	40

distance from centre /cm (+ to the right, − to the left)	30	28	26	24	22	20	18	16	14	12	10	8	6	4	2
meter reading /mV	14	26	20	11	9	3	2	13	14	22	40	28	10	9	28

Plot a graph of meter reading against distance from the centre and draw a smooth best-fit curve through the points.
Measure the distance between the maxima on either side of the central peak. This is $2\Delta x$.
Calculate a value for the wavelength of the microwaves.

Answer ~2.7 cm.

Evaluation This is not a precise way to measure the wavelength of microwaves. The main uncertainty is in the value of Δx. It is difficult to locate the two maxima precisely, and the uncertainty in fringe separation could be 2 cm in 10 cm (= 20%). The uncertainty in d will be about 0.5 cm in a distance of 10 cm (= 5%), and in D about 0.5 cm in a distance of 50 cm (= 1%). The overall uncertainty will therefore be about 26%.

Experiment 2 Making an interference pattern with light waves and using it to measure their wavelength

Apparatus

- laser (Class 2 or 3a)
- white screen with a strip of millimetre graph paper to act as a scale
- metre rule
- double slit (bought or home-made – see below)
- light sensor and datalogger (if available)

To make a double slit:

- Use a pin to draw two parallel scratches on a glass slide that has been painted with colloidal graphite. They should be clear scratches less than 0.5 mm apart.
- Or use a graphics program on a computer to draw a black-filled rectangle with two white parallel lines crossing it. Print it onto a clear plastic overhead transparency sheet.

See *Topics in Safety* 3rd Edition, ASE, 2001 (p. 118) for a full discussion of the safe use of lasers in a school laboratory.

Plan

- Position the laser about 2 m away from the screen. Put the double slit in a clip in front of the beam.
- **Tell observers to look only at the screen** and switch on the laser.
- Adjust the double slit until it lies in the path of the beam and an interference pattern is seen on the screen.
- Adjust the pattern to lie on the millimetre scale and measure the distance between as many interference fringes as possible. Use this to get the average fringe separation (Δx). Measure the distance from the double slit to the screen (D).
- Measure the separation of the centres of the slits (d). This can be done by photocopying the slits at double size again and again, until they are big enough to measure with a vernier calliper or metre rule. Alternatively, use a travelling microscope.
- If you have a datalogger with a small light sensor, set it to measure every 50 ms for 100 readings and move the sensor a known distance, steadily across the pattern. The data will allow you to plot an intensity graph of the interference pattern and obtain an average value of Δx (see the *Sample readings* overleaf).

Analysis

1. Write your results with a clear statement of what each quantity is, its unit and the estimated uncertainty.

2. Use the results to calculate a value for the wavelength of the light from the laser:

$$\lambda = \Delta x \, d/D$$

Sample readings

Double slit to screen distance = 280 cm.
Separation of slits = 0.5 mm.
A light sensor was moved steadily across the central 5 cm of the pattern and datalogger time readings were converted into distances.

position /mm	10.0	10.5	11.0	11.5	12.0	12.5	13.0	13.5	14.0	14.5	15.0
relative intensity	10.7	12.0	12.0	10.4	8.5	8.4	8.7	9.1	11.0	12.2	11.6

position /mm	15.5	16.0	16.5	17.0	17.5	18.0	18.5	19.0	19.5	20.0	20.5
relative intensity	9.1	8.3	7.5	7.4	7.4	7.5	8.0	8.5	9.0	9.5	10.4

position /mm	21.0	21.5	22.0	22.5	23.0	23.5	24.0	24.5	25.0	25.5	26.0
relative intensity	11.0	10.4	9.2	9.8	12.0	14.1	21.7	23.7	21.1	16.3	14.8

position /mm	26.5	27.0	27.5	28.0	28.5	29.0	29.5	30.0	30.5	31.0	31.5
relative intensity	16.9	27.2	31.3	34.7	36.2	35.7	34.2	22.7	21.4	27.8	36.8

position /mm	32.0	32.5	33.0	33.5	34.0	34.5	35.0	35.5	36.0	36.5	37.0
relative intensity	41.2	40.2	32.4	25.0	23.0	25.9	34.1	40.8	42.3	42.6	37.3

position /mm	37.5	38.0	38.5	39.0	39.5	40.0	40.5	41.0	41.5	42.0	42.5
relative intensity	23.1	26.9	35.5	38.5	36.9	31.0	22.6	19.9	17.5	19.4	24.6

position /mm	43.0	43.5	44.0	44.5	45.0	45.5	46.0	46.5	47.0	47.5	48.0
relative intensity	26.0	28.9	25.7	14.7	14.8	15.6	13.6	10.8	7.6	7.0	6.9

position /mm	48.5	49.0	49.5	50.0	50.5	51.0	51.5	52.0	52.5	53.0	53.5
relative intensity	7.0	7.6	7.3	7.2	8.3	8.1	6.9	7.3	7.4	6.7	6.2

position /mm	54.0	54.5	55.0	55.5	56.0	56.5	57.0	57.5	58.0	58.5	59.0	59.5
relative intensity	6.3	6.5	6.5	6.3	6.0	6.0	5.9	6.6	7.2	7.4	7.6	7.8

Plot a graph of the sample readings. It will show the interference fringes inside an overall diffraction pattern envelope.
Obtain an average value of fringe separation and hence a value for the wavelength of the laser light.

Answer ~643 nm.

Evaluation The uncertainty in D may be 1 cm in 200 cm (= 0.5%).

The sample readings show that 5 fringes have a separation of 18 mm. The fringes do not have sharp edges and the uncertainty in this measurement could be ±2 mm. This gives a value of 3.6 ± 0.4 mm for Δx (~11% uncertainty).

The slit separation carries a large percentage uncertainty because it is so small. For the sample readings the double slit was enlarged 8 times and the separation was then 4 mm with an uncertainty in its measurement of about 0.5 mm (~8% uncertainty). Add to this the inaccuracies of the photocopier and the overall uncertainty in d could be 14%.

The overall uncertainty in the wavelength would then be 11% + 14% + 0.5%, giving a total of around 25%. This is an example of an accurate but imprecise experiment!

Improving the plan

- Use a datalogger to measure and plot the light intensity across the pattern, as already suggested. The readings make it much easier to identify the maxima of the interference fringes and obtain a more precise value for the fringe separation.
- Use a slide projector to form a highly magnified image of the double slit on a wall. The edges will not be well defined at high magnifications but it should increase the accuracy of the measurement of the slit separation. The magnification will be:

  screen to lens distance/lens to slide distance

12.6 The photoelectric effect

Computer simulation: Photon

(There are no lab experiments for this topic.)

Computer simulation Photon

Aim **To investigate the photoelectric effect.**

Apparatus

- computer running the program 'Photon' from the CD

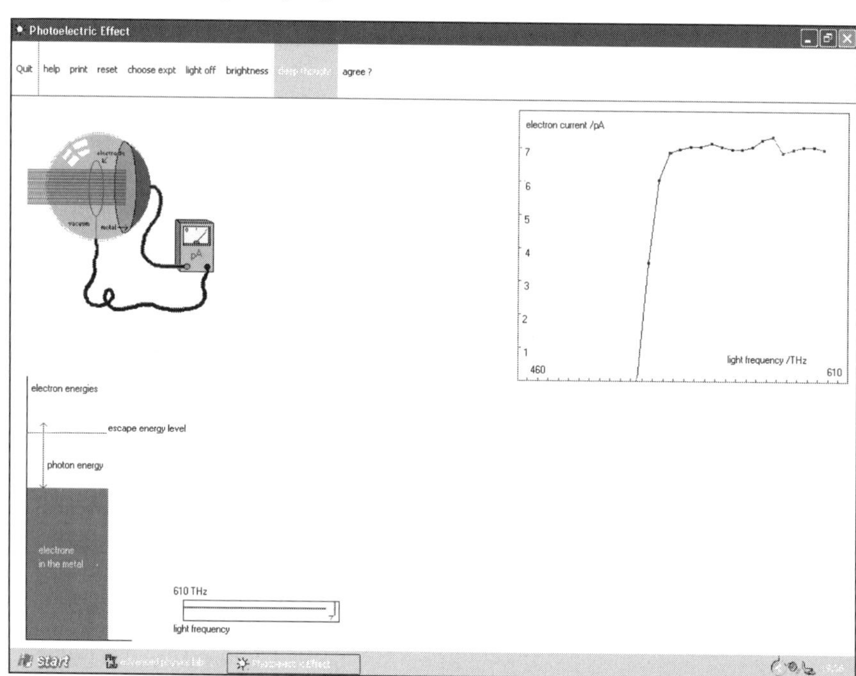

Plan

- Start with Experiment A. Switch on the light and note that a current flows across the vacuum in the cell.
- Experiment B (shown above) allows you to alter the frequency (colour) of the light. Notice that a current flows only after the frequency reaches a certain level.
- Use Experiment C to measure the energy of the electrons released. A negative potential is applied to the cell until the current is cut off. The size of this stopping pd is a measure of the maximum energy of the electrons. Plot the maximum energy of electrons released by light of different frequencies.
- Fill in the 'agree?' dialogue box.

Analysis Print out your readings, graphs and conclusions, and the theory from the 'help' notes. Add your own notes to produce a report on the topic.

13 Optics

13.1 The refraction of light

Preliminary work: Observing effects caused by the refraction of light
Experiment: Measuring the refractive index of a glass or Perspex block

Preliminary work

Observing effects caused by the refraction of light

Apparatus

- drinking mug
- coin
- jug of water
- glass block
- pin and drawing pin
- piece of cork
- striped 'bendy' plastic drinking straw
- large tank of water (e.g. fish tank)

Plan Try some of these well known effects for yourself.

- Put a coin in a deep mug and position your eye so that the coin just can't be seen. Hold your head steady and carefully fill the mug with water. The edge of the coin should come into sight.
- Put a plastic straw into a tank of water and notice that it appears to be bent, especially if you look along its length. Make a bend in the straw before you put it in the water. Then, while it is in the water, adjust the angle of the bend until it looks straight. A straw with stripes along it works best.
- Look through a block of glass at a pin stuck in a piece of cork at its far end. To judge where its image in the block appears to be, move a drawing pin on the top surface of the glass block until the drawing pin and the image of the pin appear to be the same distance away, i.e. there is no parallax between them.

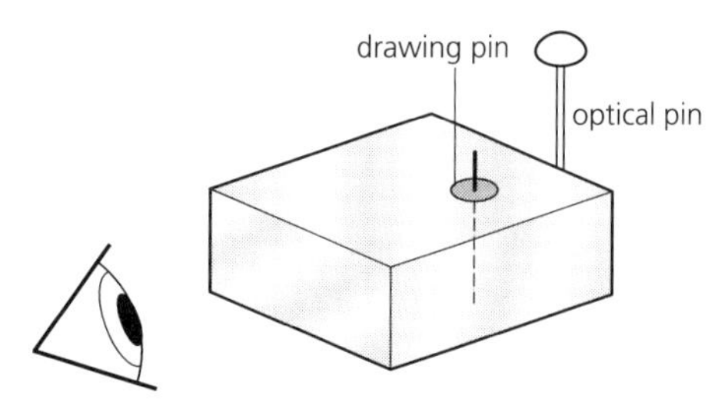

Figure 13.1

- Next time you are standing in clear still water, look at your feet and notice how refraction seems to shorten your legs.

Experiment: Measuring the refractive index of a glass or Perspex block

Apparatus

- glass or Perspex block (rectangular or semi-circular)
- ray box and power supply
- 5 sheets of 1 mm squared graph paper

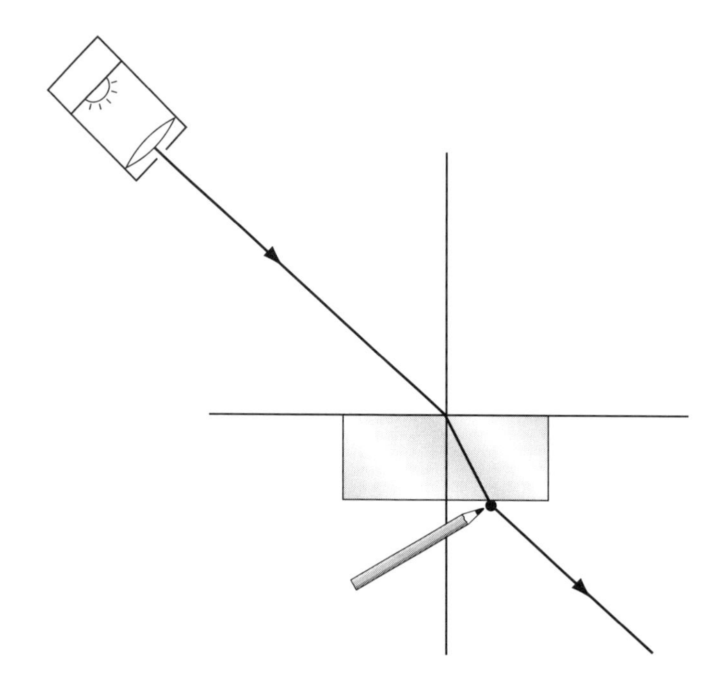

Figure 13.2

Plan

- Draw two intersecting horizontal and vertical lines in the middle of the graph paper along the darker rulings.
- Draw a line at an angle, to meet the perpendicular lines at the place where they intersect. This will be the direction of your incident ray (see Figure 13.2).
- Place the block carefully along the horizontal line and shine a ray of light from the ray box along the incident ray line. Make the ray from the box as narrow and sharp as possible, and use the edge of it as your guide for positioning.
- Look into the block from the other side and mark the bright spot where the ray leaves the block (see Figure 13.2).
- Remove the block and draw in the refracted ray in the glass.
- Repeat with a clean piece of paper. Obtain 5 drawings of incident and refracted rays for a wide range of angles of incidence.

Skill level (Implementing)

A: I made a long, fine beam of light with a ray box slit and lens. I constructed two axes on mm graph paper and lay the block along one of them. I shone the beam onto the block and located the refracted beam. I repeated with a wide range of angles. I used a separate sheet of graph paper for each angle.

All but one of the above = B; all but two = C; all but three = D; all but four = E.

Analysis

1 For each drawing, place the point of a compass at the point of incidence and draw the largest circle you can that cuts the two rays. Use the millimetre scale of the graph paper to measure the distances x and y (see Figure 13.3).
2 Make a table for the values of x, y and y/x.

Simple trigonometry shows that:

$\sin i/\sin r = y/x$

and so y/x is the refractive index of the material of the block.

3 Work out an average value of the refractive index and the percentage uncertainty.

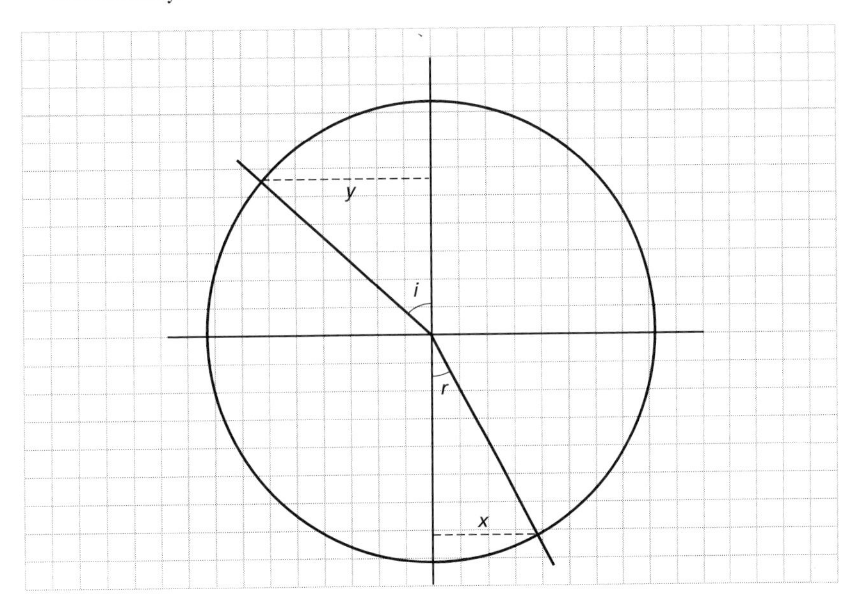

Figure 13.3

Sample readings

y /mm	93.0	88.0	82.0	76.0	66.0	60.0	48.0	41.0	
x /mm	62.0	58.0	54.0	49.0	43.0	40.0	31.0	26.0	
refractive index	1.50	1.52	1.52	1.55	1.54	1.50	1.55	1.58*	average 1.53
numerical difference from the average	0.03	0.01	0.01	0.02	0.01	0.03	0.02		std 0.03

*considered to be anomalous so not used for calculating average

Plot a graph of y against x for the sample data. Draw the best-fit straight line and calculate a value for refractive index from the gradient. Decide whether the line goes through the origin and thus whether the light rays follow Snell's law.

Answers $n \approx 1.51$; yes.

Evaluation All the uncertainties come from the drawings. Judging the position of the edge of the light beam and where it emerges from the glass block introduces an uncertainty in the position of the lines. Positioning the block, constructing the lines and reading distances from the graph paper will add to the uncertainties in x and y. These can be evaluated from the numerical differences of the values from the average or from the standard deviation. The sample readings have an uncertainty of about 2%.

If you had tried to measure the angles of incidence and refraction with a protractor and then looked up the sines of those angles, the uncertainties would have been much greater, especially when the angles were small.

Improving the plan

- Try to improve the light beam by adapting the ray box. Use a lens to produce a long fine bright beam of light and a plastic filter to colour it.
- Darken the room to help see exactly where the beam enters and leaves the block.

13.2 Total internal reflection

Preliminary work: Observing total internal reflection
Experiment: Measuring the critical angle for glass or Perspex

Preliminary work

Observing total internal reflection

Apparatus

- glass prism (45°–45°–90°)
- flat Perspex prism (45°–45°–90°)
- optical fibre
- ray box and power supply
- torch

Plan

Observe these well known examples of the reflection of light from the inside surfaces of plastic and glass.

- The sides of a 45° glass prism can be used as reflecting surfaces to see sideways or backwards. Lay the prism on the words of a book to see how well they are reflected and whether the image is reversed.

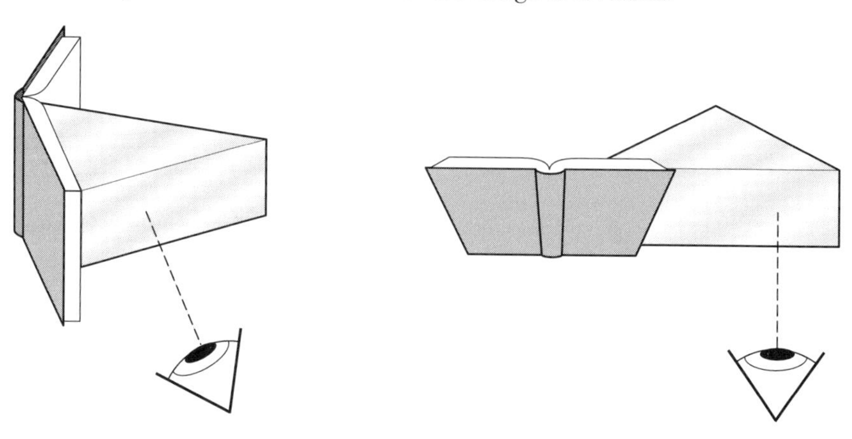

Figure 13.4

- Use the flat Perspex prism and ray box to make a beam of light reflect through 90° and 180° (see Figure 13.5).

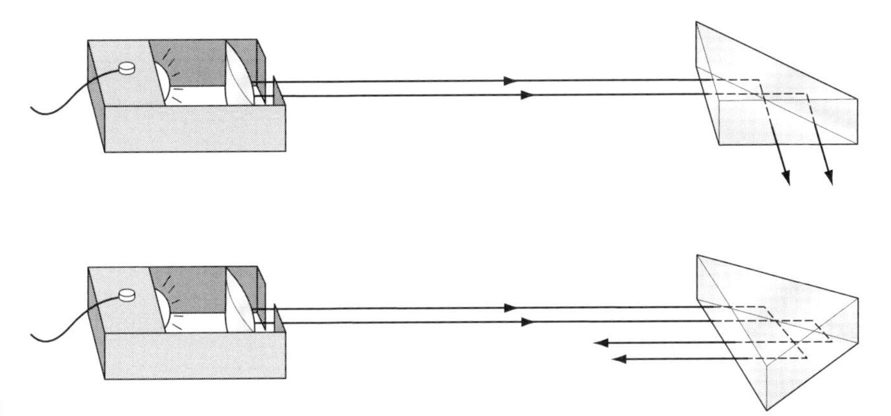

Figure 13.5

- Use the optical fibre and torch to direct light into a remote dark place.

Experiment

Measuring the critical angle for glass or Perspex

Apparatus

- glass or Perspex semi-circular block
- ray box and power supply
- coloured plastic filter
- 6 sheets of 1 mm squared graph paper

Plan

- Draw vertical and horizontal axes in the middle of a sheet of graph paper along the darker lines.
- Measure the length of the straight edge of the block and place the centre of that edge on the origin of your axes.
- Use the ray box and filter to get a thin sharp beam of coloured light.
- Aim the light through the curved surface of the block at the centre of the straight edge. It will hit the curved surface at 90° and pass straight through to the straight edge.
- Turn the graph paper and block until the beam of light just emerges at the centre of the straight edge and skims out along the straight surface (see Figure 13.6).

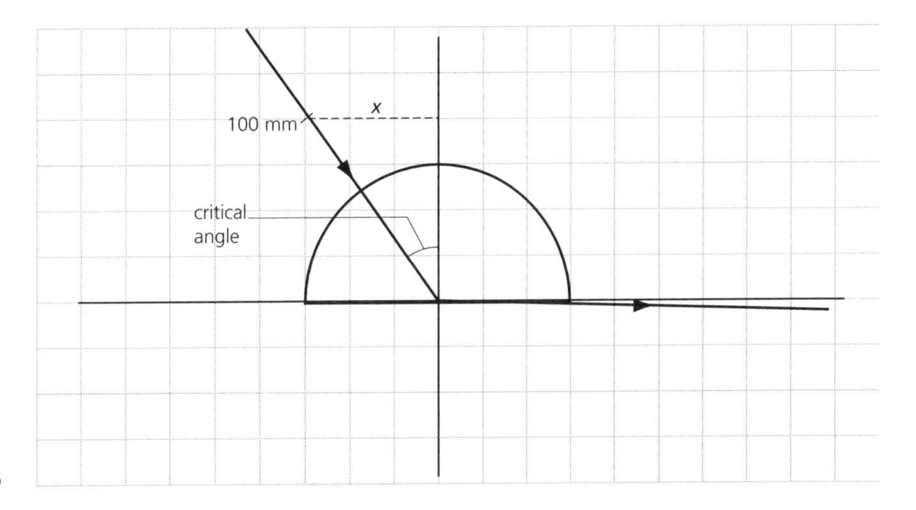

Figure 13.6

- Draw dots along the edge of the incident beam to mark its position. This is known as the 'critical' ray.
- Repeat the same procedure 5 times on separate pieces of graph paper.

Analysis

1. Join the dots to mark the direction of the critical ray and make a mark on it, 100 mm from the origin of the axes. Read the distance x mm from this mark to the y-axis (see Figure 13.6).
2. Calculate the sine of the critical angle ($x/100$) hence obtain a value for the critical angle, $\sin^{-1}(x/100)$.
3. Do the same for your other drawings and work out the average as your best value for the critical angle.

Evaluation Uncertainties are introduced by the width of the light beam and the difficulty of deciding when the ray is just critical. The beam is dispersed when it leaves the glass and the component colours are critical at different angles. Added to these are uncertainties introduced by drawing and measuring from the graph paper. You can get an estimate of the likely uncertainty from the numerical difference of each of your repeat readings from the average. Take the uncertainty to be half the largest difference.

14 Thermal properties

14.1 Heating materials

Experiment: Measuring the specific heat capacity of a metal block
Full investigation: Comparing the specific heat capacities of cooking oil and water

Experiment — Measuring the specific heat capacity of a metal block

Apparatus

- solid metal block with holes for a heater and a thermometer
- low voltage heater to fit into the block
- electronic or mercury thermometer (0–100 °C)
- oil (e.g. cooking oil)
- balance
- low voltage power supply (smoothed if possible)
- ammeter (0–5 A)
- voltmeter (0–15 V)
- connecting leads
- stop clock or watch
- datalogger (optional)

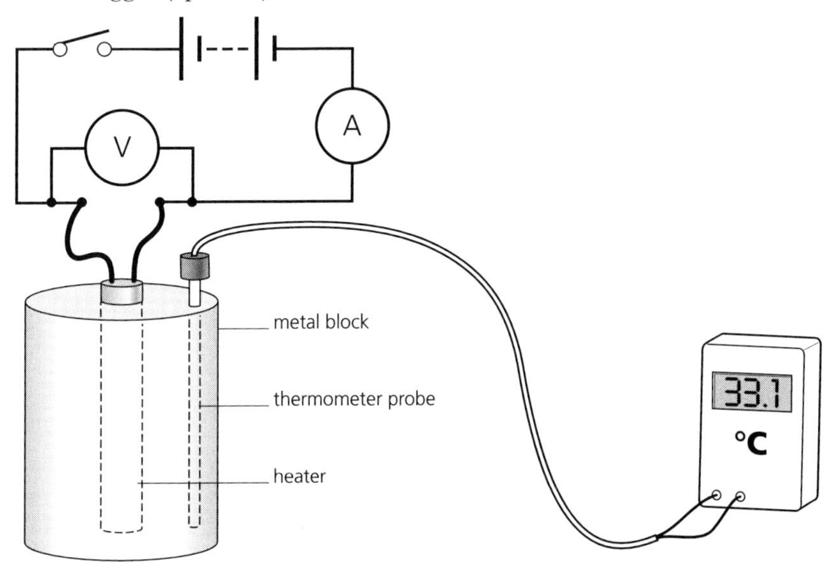

Figure 14.1

Plan

- Connect the heater and ammeter in series to the power supply.
- Connect the voltmeter across the heater.
- Adjust the voltage of the power supply to that required by the heater. Switch on briefly to check that the meters read correctly.
- Put the heater and thermometer into the metal block. Put a little oil into the thermometer hole to ensure good thermal contact. Record the starting temperature.
- Switch on the heater for 10 minutes, checking the current and pd at intervals and recording them if they change.

- Record the final temperature as the highest reached by the metal. Leave the apparatus to cool for exactly 5 minutes after switching off. If there is a drop in temperature, add this drop to the final temperature to get the 'corrected' final temperature (see *Evaluation* opposite).
- Measure the mass of the block.

Analysis

1 Draw up tables with the following entries for your results:

Type of metal
Mass (m) /kg
Starting temperature (θ_1) /°C
Final temperature (θ_2) /°C
Corrected final temperature (θ_3) /°C
Temperature rise ($\theta_3 - \theta_1$) /°C

Current (I) /A
pd (V) /V
Time of heating (t) /s
Electrical energy supplied (VIt) /J

2 Enter your readings and calculate the specific heat capacity.

Specific heat capacity $c = VIt/m(\theta_3 - \theta_1)$

Sample readings

Type of metal: aluminium
Mass /kg = 1.00
Current /A = 1.66
pd /V = 9.60
Heating time = 12 minutes

The temperature changes are shown in the graph, produced by an electronic thermometer and datalogger.

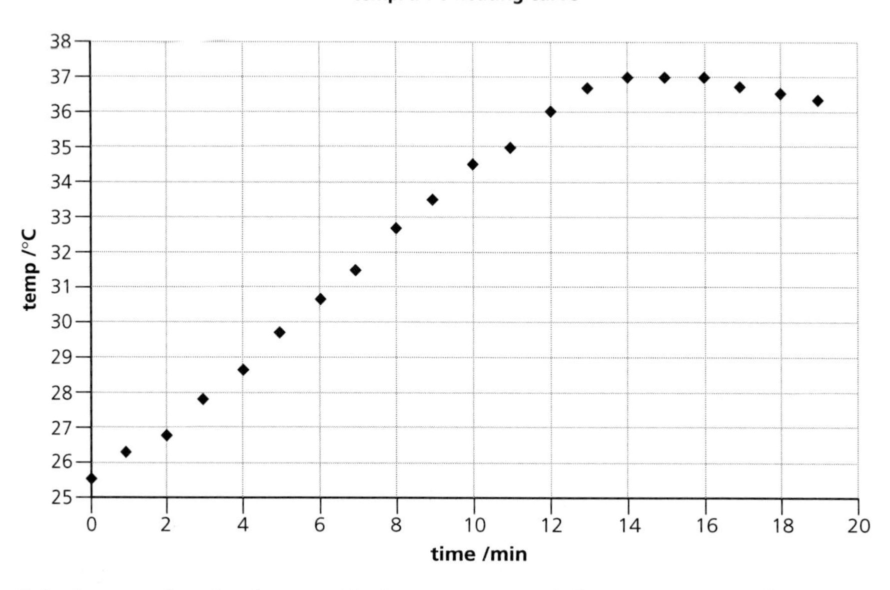

Figure 14.2

Calculate a value for the specific heat capacity of aluminium using the sample readings.

Answer 972 J kg^{-1} K^{-1}.

Evaluation If the power supply gives a smooth output, the current and voltage readings should be as precise as the calibration given by the manufacturer – typically 2%. The mass should have only a small uncertainty, of probably less than 1%. The temperatures can be read to about 0.5 °C but when the readings are subtracted a smaller quantity is obtained which has a larger % uncertainty. This could be 1 °C in 15 °C (≈ 7%).

The cooling correction suggested in the *Plan* lets the block cool at its maximum rate for half of the heating time and equates this to the loss at an average rate while it is being heated. No insulation is used because this would take an unknown amount of energy from the heater. The *Sample readings* show that only a small correction is necessary and that the temperature loss due to cooling is small.

The total uncertainty due to all readings would then be about 11%.

The heater itself is a cause of inaccuracy. It takes energy from the supply as it warms up and gives a result for the specific heat capacity that is too high.

Improving the plan

- Allow for the energy taken by the heater, by estimating its specific heat capacity and multiplying by its mass and the temperature rise. Deduct this quantity from the total energy supplied.
- Use a temperature probe and datalogger to plot the temperature changes on a computer, as was done for the *Sample readings*. Read the final temperature and cooling correction from the graph.
- Use a longer heating time. The temperature rise would be larger and its % uncertainty lower (but the heating loss and its uncertainty would be greater).

Full investigation

Comparing the specific heat capacities of cooking oil and water

Apparatus

- electronic or mercury thermometer (0–100 °C)
- balance
- low voltage power supply (smoothed if possible)
- ammeter (0–5 A)
- voltmeter (0–10 V)
- connecting leads
- stop clock or watch
- 1 m constantan resistance wire (SWG 26 or similar)
- single-strand insulated copper wire
- 2 terminal blocks
- copper container
- cooking oil and water
- stirrer
- datalogger (optional)

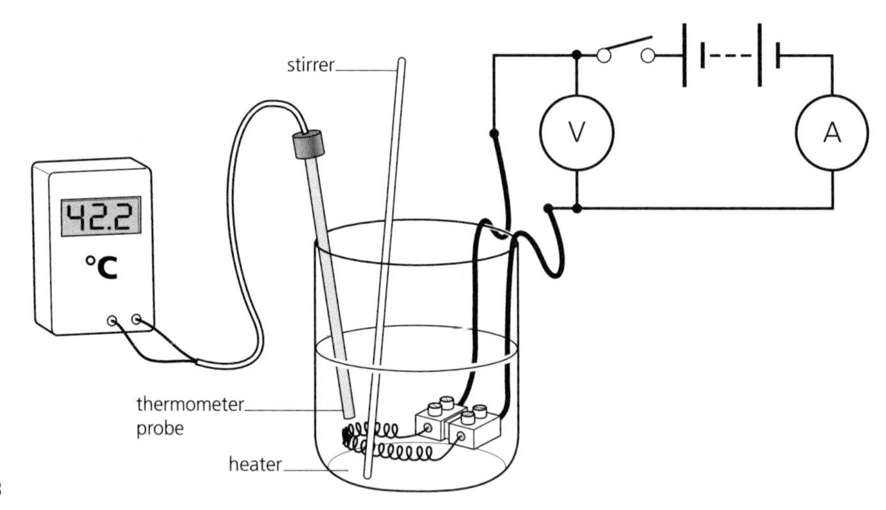

Figure 14.3

Plan

- Make a heater by winding the resistance wire onto a pencil. Remove the pencil and connect the ends of the wire to the terminal blocks. Make two short leads from the single-strand copper wire and screw them to the other sides of the terminal block.
- Build a series circuit using the heater, ammeter and power supply. Connect the voltmeter in parallel across the heater. Switch on very briefly and turn up the voltage until a suitable current (~2 A) flows through the heater.
- Measure the mass of the copper container, fill it half-full with cooking oil and measure its mass again.
- Immerse the heater in the oil, making sure all of its coils are covered and that they are not touching each other or the container. Use the stiff copper leads to position the coil.
- Take the starting temperature of the oil, switch on the current and heat the oil for 10 minutes.
- Stir very thoroughly and measure the highest temperature reached by the oil.
- Let the apparatus cool for half the heating time and add any drop in temperature to the highest temperature to find the corrected final temperature.
- Repeat the whole procedure with water.

Skill level (Implementing)

A: I made a heater with leads. I built a circuit that could measure the current and pd used by the heater. I heated the oil for a measured time and found the rise in temperature. I measured the cooling correction. I repeated the experiment with water.

All but one of the above = B; all but two = C; all but three = D; all but four = E.

Analysis

1. Put your results into a table similar to that of the previous experiment, except you will need extra entries for the mass of the container with and without oil, and the mass of oil.
2. To find the specific heat capacity, you will have to work out the energy supplied electrically, the corrected temperature and the mass of the oil. (See the previous experiment for the calculation, p. 170.)
3. For a more accurate result, calculate the energy taken to warm the container and subtract it from the total to find the energy used to heat the oil (see *Improving accuracy* below).

Sample readings

	cooking oil	**water**
mass of copper container /g	102.43	102.43
mass of container + oil or water /g	372.43	312.82
mass of oil or water /g		
pd /V	6.4	6.4
current /A	2.0	2.0
heating time /minutes	10	10
cooling time /minutes	5	5
specific heat capacity of copper /$J\ kg^{-1}\ K^{-1}$	380	380
initial temperature /°C	25.2	25.2
maximum temperature reached /°C	42.2	33.1
cooling correction /°C (using an electronic thermometer and datalogger)	1.0	0.5

Using these sample readings, calculate the specific heat capacities of cooking oil and water following the steps listed in the *Analysis*.

Answers

cooking oil: 1440 $J\ kg^{-1}\ K^{-1}$; water 4170 $J\ kg^{-1}\ K^{-1}$.

Improving accuracy

There is an extra systematic error with such an experiment with a liquid because about 5% to 15% of the energy supplied is taken to warm up its container and the stirrer. Accuracy can be improved by calculating the exact amount of energy taken by the container (= mass × specific heat capacity × temperature rise) and deducting this quantity from the energy supplied by the heater, to get the energy supplied to the liquid.

Polishing the copper container with metal polish until it shines will reduce the amount of heat lost by radiation.

14.2 Latent heat

Preliminary work: Observing the changes that take place when water is heated until it boils
Experiment: Measuring the specific latent heat (of vaporisation) of water
Full investigation: Measuring the specific latent heat (of vaporisation) of water more accurately

Preliminary work — Observing the changes that take place when water is heated until it boils

Apparatus

- beaker of water
- Bunsen burner, tripod, gauze, bench mat
- electronic or mercury thermometer (0–110 °C)
- safety glasses

Plan

- Fill the beaker two-thirds full of cold tap water, light the Bunsen beneath it and record all that you see, including the changing temperature, until the water is boiling freely.

Analysis Put your observations into a table like this:

Observation	Temperature
Mist appears on the outside of the beaker.	
Small bubbles rise from the bottom of the beaker to the top.	
'Wisps' appear above the water surface.	
Large bubbles appear and disappear on the bottom of the beaker.	
Large bubbles get to the surface where they burst and whip the surface into a turmoil.	
The water boils steadily but gets no hotter.	

Experiment — Measuring the specific latent heat (of vaporisation) of water

Apparatus

- old plastic electric kettle
- measuring cylinder (100 cm^3)
- thermometer
- stop clock or watch
- safety glasses and oven gloves

Plan **Extreme care is needed during this experiment.** Wear safety glasses, and remember that water vapour is invisible and contains a lot more internal energy than boiling water. Arrange the apparatus so that you can operate it from a distance and there is no danger of you, or others, being scalded by the steam or boiling water.

- Carefully measure out 1 kg (=1 litre) of water into the kettle and measure its temperature (θ).

- Leave the lid off the kettle, switch on and start the stop clock. Note the time (t) it takes the water to boil.
- Keep the clock running and let the water boil for 5 minutes. (You may have to over-ride the automatic switch on the kettle.)
- Carefully pour the water that remains into the measuring cylinder, to find out the mass of water that has been boiled away.

Analysis

1 Use the time it takes the water to boil and the specific heat capacity of water to estimate the power supplied to the water by the heating element.

Specific heat capacity of water = 4180 J kg^{-1} K^{-1}
Energy supplied to the water = $1 \times 4180 \times (100 - \theta)$ J
Average power (P) supplied to the water = $4180\,(100 - \theta)/t$ W

2 Record all your readings and calculations as follows:

Mass of water = 1 kg
Time to reach boiling point (t) =
Temperature of water at start (θ) =
Rate at which energy is supplied to water by heater (P) =
Mass of water turned into vapour (m) =
Boiling time = 600 s
Energy supplied to water during boiling ($600 \times P$) =
Energy needed per kg to turn water into vapour at its boiling point ($600P/m$) =

Sample readings

Mass of water = 1 kg
Starting temperature = 24.0 °C
Time to reach boiling point = 3 minutes 23 s
Boiling time = 5 minutes
Mass of water boiled away = 230 g

Calculate the specific latent heat of vaporisation from these sample readings.

Answer 2.1 MJ kg^{-1}.

Evaluation

The method of calculating the average power supplied to the water (from specific heat capacity and temperature rise) is more precise than using the given power of the heating element. However, the energy supplied will not be an accurate figure, because some is used to heat the element and the body of the kettle. A second source of inaccuracy is the heat lost from the hot kettle by convection and radiation. Energy has to be supplied to make up for this, and less vapour is produced than if there had been no losses. Also, some of the vapour condenses and falls back into the water.

A 100 cm^3 measuring cylinder is suggested for measuring the water because, although it is laborious to use, it measures volume precisely. The quantity with the largest uncertainty is the mass of water that turns into vapour (m). It is obtained from subtracting two measurements, each uncertain by at least 4 g, and so has an uncertainty of 8 g in 200 g (= 4%). The uncertainty in the calculated power supplied to the water comes from measurements of the initial mass of water (4 g or ~0.4%), the temperature rise (1 °C or ~2%) and the time to reach boiling (20 s or ~10%). These would give an overall uncertainty in the value of the specific latent heat of ~17%.

Full investigation

Measuring the specific latent heat (of vaporisation) of water more accurately

Apparatus As for the previous experiment, plus:

- digital balance
- electronic thermometer probe and datalogger

Figure 14.4

Plan Take the same safety precautions as in the previous experiment: see p. 174. Follow the same method, except:

- Use an electronic balance to measure the mass of water before and after boiling.
- Use an electronic thermometer and a datalogger to plot the temperature of the water as it is brought to the boil.
- Find the power supplied to the water near to the boiling point when losses will be similar to those during boiling.
- Boil the water for longer than 5 minutes to increase the quantity turned into vapour.

Analysis Power supplied to the water:

P = mass of water × specific heat capacity
× gradient of the temperature/time curve close to boiling point

Energy supplied to the water during boiling:

$E = P \times$ boiling time

Specific latent heat of water:

$L = E$/mass of steam produced

Use these equations to calculate the specific latent heat of vaporisation of water.

Sample readings

time after switch-on /s	150	155	160	165	170	175	180	185	190	195	200	205	210	215	220	225
temperature /°C	73.8	76.6	78.8	81.0	83.0	86.0	90.0	93.4	96.0	97.8	99.4	100.4	101.0	101.4	101.6	101.8

The kettle was switched on at time 0 s and off at time 720 s.
The mass of water started at 1000 g and finished at 564 g.

Plot the temperature/time graph for the sample readings and estimate the gradient just before the water boils.
Calculate the power given to the water by the heater.

Answer ~2260 W.

Estimate the boiling time and the mass of vapour produced.
Calculate the specific latent heat of vaporisation of water.

Answer ~2.6 MJ kg^{-1} K^{-1}.

Evaluation Using a balance to measure the loss in mass of the water is more precise than pouring the water into a measuring cylinder. The uncertainty may be about 2 g in 400 g. However, water turns to vapour before it boils, and some steam re-condenses inside the kettle during boiling. This could involve an extra 2 g, giving a total uncertainty in the mass of vapour of 4 g (~1%).

Estimating the gradient of the temperature/time graph to calculate the power will introduce an uncertainty of ~5%.

The moment of boiling is not precise and introduces an uncertainty of ~20 s into the time of boiling. Lengthening the boiling time will reduce the significance of this to say 20 s in 500 s (~4%).

Overall, the uncertainty of the modified method will be ~10%. As with all experiments, there is a limit placed by the design on the accuracy that can be achieved.

14.3 Absolute zero

Computer simulation: Gas
Full investigation: What is the absolute zero of temperature?

Computer simulation Gas

Aim **To measure the expansion of an ideal gas when it is heated at constant pressure.**

Apparatus

- computer running the program 'Gas' from the CD

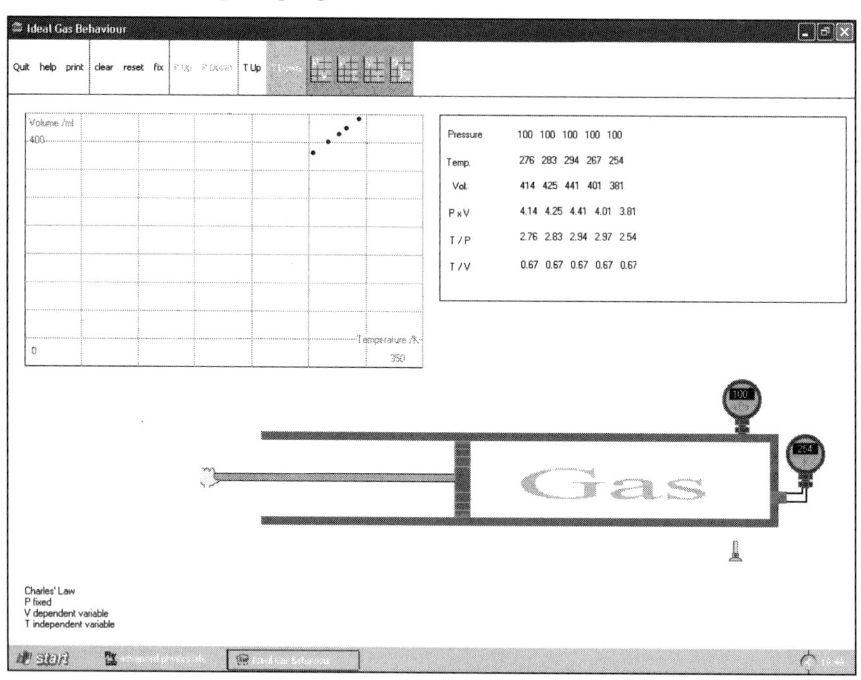

Plan The simulation program can be used to experiment with all three gas laws. For this particular investigation of the expansion and contraction of a gas when heated, volume and temperature are varied and pressure is kept constant.

- Click on 'fix'. In the box that comes up, tick 'Fix pressure' and leave the temperature and volume tick boxes clear.
- Click 'T Up', and a small Bunsen burner starts to heat the gas in the cylinder. The thermometer reading rises and the piston moves out as the gas expands. When you click 'T Up' again, the heating is stopped and the readings are recorded.
- Click on the 'V/T' graph button to see a plot of volume against temperature in kelvin.
- Continue to heat (up to 300 K) and then cool ('T Down') the gas until you have enough points on your graph to show how volume and temperature are related.
- Print a copy of the results.

Analysis Draw the line of best fit on the graph and write a conclusion to go with the readings.

Full investigation

What is the absolute zero of temperature?

Apparatus

- glass capillary tube, length 30 cm, sealed at one end and containing a thread of concentrated sulfuric acid (this acts as a moving piston that traps and dries a column of air)
- deep beaker
- electronic or mercury thermometer (0–110 °C)
- Bunsen burner, tripod and gauze
- half-metre rule
- clamp and stand
- small rubber bands
- stirrer
- crushed ice
- safety glasses

The tube should be prepared by a technician wearing eye protection. Seal a 1 mm internal diameter capillary tube at one end with a glass-bending Bunsen or gas burner. Heat the glass along its length to well above 100 °C and invert its open end into a small beaker of concentrated sulfuric acid. Allow the tube to suck in a 2 cm length of acid. Remove the tube from the acid and allow it to cool. Carefully mop up excess acid with a paper towel.

Plan **Take care when doing this experiment**. Make sure the beaker is stable, and use a clamp and stand to hold the metre rule. **Wear safety glasses** throughout the heating process and be aware, if acid escapes from the end of the capillary tube, that it is very corrosive.

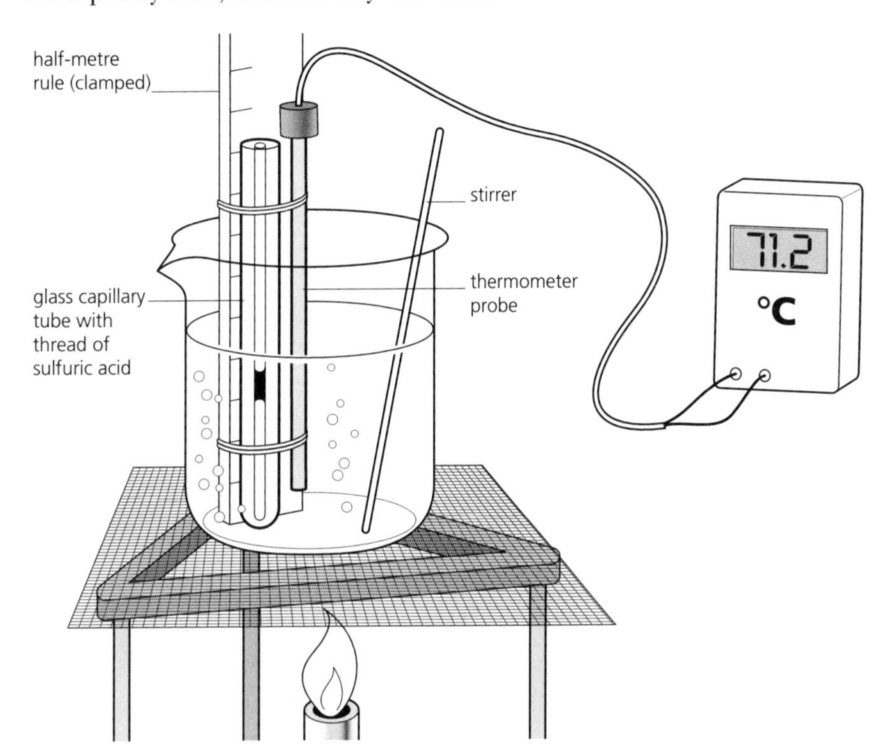

Figure 14.5

- Put tap water in the beaker (no more than three-quarters full), add crushed ice and leave the water to cool.
- Fix the capillary tube, thermometer probe and half-metre rule together with the rubber bands and place in the water. The air column should be completely submerged in the water. Put the beaker on the tripod and gauze.
- Stir the water thoroughly and, when the temperature has stabilised, read the length of the air column trapped by the acid thread and record the temperature.
- Warm the water by about 10 °C, remove the Bunsen and stir the water carefully but thoroughly. Read the temperature when it becomes steady, and the length of the air column.
- Continue heating, taking readings about every 10 °C until the water boils.

Skill level (Implementing)

A: I cooled the water before I started and made sure all the ice had melted. I took readings of length and temperature from just above the ice point until boiling started. I removed the Bunsen, stirred and waited for the temperature to become steady before each reading. I clamped the tube and rule securely and kept thermometer leads away from the flame. I wore safety glasses throughout and there were no near accidents with boiling water or Bunsen flames.

All but one of the above = B; all but two = C; all but three = D; all but four = E.

Analysis

1. Record your length and temperature readings in a table.
2. Plot a graph of length (y-axis) against temperature (x-axis). The x-axis scale needs to go from −300 °C to +100 °C and the y-axis scale from zero.
3. Draw the best-fit line and extrapolate it back until it cuts the x-axis. Read this temperature (where the volume of the air would have become zero) – it is your estimate of the absolute zero of temperature.

Sample readings

temperature /°C	0.8	10.8	22.3	35.7	44.5	51.9	61.7	72.4	83.2	91.1	100.4
length /mm	69	72	74	77	79	81	83	86	88	90	93

Plot the graph described in the *Analysis* above for these sample readings. Read off the intercept on the x-axis.

Answer ~ −282 °C.

Plot a second graph with the x-axis from 0 to 100 °C and the y-axis from 60 to 100 mm. Find the gradient of the straight part of the line.

Answer ~0.238 mm /°C.

This is the amount the length increases or decreases for a temperature change of 1 °C. Read off the length of the air column at 0 °C and calculate how many degrees it would have to be cooled to contract to zero length.

Answer ~ −289 °C.

Evaluation The temperature readings will have uncertainties, because the water throughout the length of the air column will not be at a uniform temperature. The temperature of the trapped air could differ by about 2 °C.

The length of the air column is being read from some distance through glass and water, which could introduce an error of 2 mm.

The chief cause of error, however, is in the design of the experiment which requires the best-fit line on the graph to be extrapolated backwards. A small difference in the choice of the line produces a large difference in the intercept.

The second method, suggested for analysing the *Sample readings*, is a better way of calculating the result. A temperature scale from 0 °C to 100 °C makes it easier to choose a best-fit line, especially if the readings are taken over this whole range.

14.4 Measuring temperature

Experiment: Using a copper resistance thermometer to measure room temperature

Experiment

Using a copper resistance thermometer to measure room temperature

Apparatus

- solenoid made of very fine copper wire with a resistance of several hundred ohms (e.g. sealed reed relay solenoid from Maplin)
- multimeter (set on its 2 kΩ resistance range)
- pure crushed ice in a filter funnel
- beaker
- electric kettle to boil water
- safety glasses

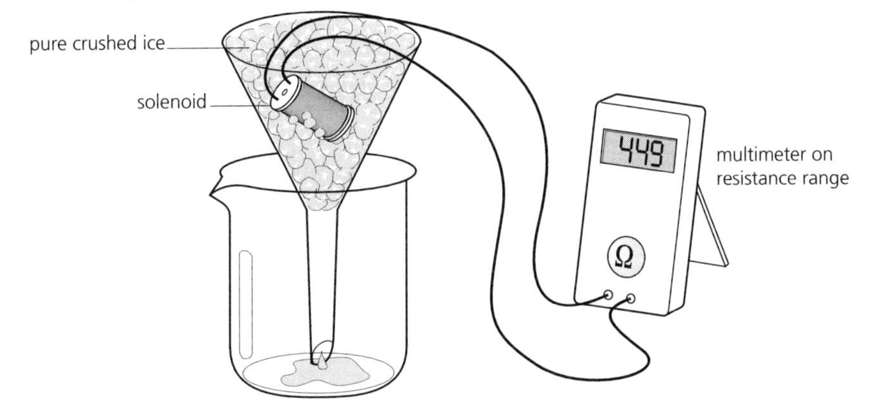

Figure 14.6

Plan

- Connect the solenoid to the multimeter set on its 2 kΩ resistance range.
- Completely cover the solenoid with crushed ice in the funnel and leave it until the ice is melting freely. Once the resistance reading is steady, record its value at the ice point (R_0).
- Put the solenoid into the water in the kettle, and boil the water. Wear safety glasses. When the multimeter reading is steady, record the resistance at the steam point (R_{100}).
- Leave the solenoid in the air of the room and again wait for the resistance reading to stop changing. Read the resistance (R_θ) at the room temperature (θ).

Analysis Write down the three resistance values, R_0, R_{100} and R_θ.

The thermometer measures temperature on the 'copper resistance thermometer scale'. This assumes that the resistance of copper increases uniformly with temperature between 0 °C and 100 °C. A temperature θ between 0 °C and 100 °C can then be calculated by using ratios, or the following formula:

$$\theta/100 = (R_\theta - R_0)/(R_{100} - R_0)$$

It is an empirical temperature scale and so the result should be given in degrees centigrade.

Sample readings

Resistance at ice point /Ω = 444
Resistance at steam point /Ω = 631
Resistance at room temperature /Ω = 491

Calculate room temperature from these figures.

Answer 25.1 °C.

Evaluation The melting ice and boiling water may not be pure or at standard pressure, so their supposed 100 °C difference could have an uncertainty of 1 °C (= 1%).

The copper coil may not have reached room temperature when its resistance was read, and last-digit fluctuations of the meter could give a total uncertainty in the resistance values of about 2 Ω.

A large % uncertainty arises when the resistance values are subtracted. The uncertainties add and become a larger proportion of a smaller result. For the sample readings:

$491 - 444 = 47 \pm 4\ (\approx 9\%)$
$631 - 444 = 187 \pm 4\ (\approx 2\%)$

The 47 and 187 are then divided, giving an uncertainty of ~11%. The result is multiplied by 100 ± 1%, giving an uncertainty of 12% in the final result.

15 Radioactivity

15.1 Radiation from radioactive sources

Preliminary work: Measuring background radiation
Experiment: Identifying the types of radiation coming from a radioactive source
Computer simulation: Identifying radiation

Local Education Authority regulations on the use of sealed radioactive sources should be read and adhered to in the practical work in this topic.

Preliminary work

Measuring background radiation

Apparatus

- Geiger–Müller (GM) tube
- scaler with power supply for the GM tube
- loudspeaker (optional)
- stop clock or watch

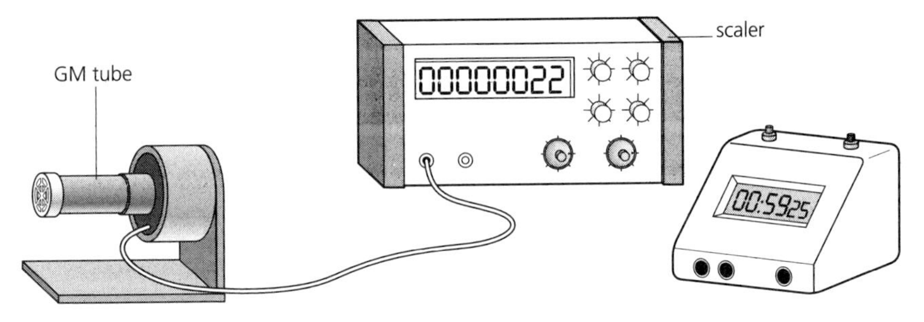

Figure 15.1

Plan

- Plug the GM tube into the scaler and adjust the voltage to its recommended level. The scaler should start to count slowly, and clicks will be heard if a loudspeaker is connected.
- Check that there are no radioactive sources nearby.
- Zero the counter and measure the count registered in 1 minute.
- Repeat 10 times.

Analysis

Put the readings into a table and find the average background count in counts/minute and the standard deviation.

Sample readings

In a school laboratory

counts/minute	12	22	19	14	15	9	11	17	14	15

Calculate the average background count for the sample readings and the standard deviation.

Answers

15 minute^{-1}; 3.6 minute^{-1}.

Evaluation

The variations of the readings are not due to errors; they are due to the random nature of radiation from radioactive sources. The GM tube will only collect a sample of the radiation that is in the room, and not all radiation entering the tube will register a pulse. These factors however are constant, and change in count rate is due to the fact that different numbers of particles enter the tube. When a GM tube is used to monitor a radioactive source, the count rate represents a fixed proportion of the source's activity and not the activity itself.

Experiment: Identifying the types of radiation coming from a radioactive source

Apparatus As for the Preliminary work, plus:

- sealed radioactive source in a holder
- long tweezers for handling the source
- absorbers for testing the radiation:
  – sheet of paper
  – aluminium plate about 3 mm thick
  – block of lead about 3 cm thick

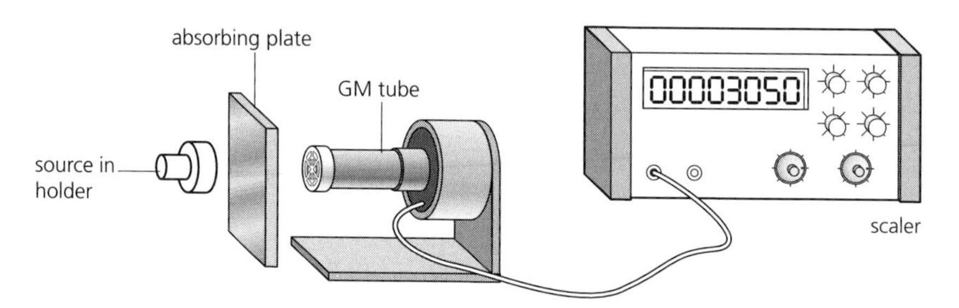

Figure 15.2

Plan **Take great care when handling radioactive materials.** Only use sealed sources approved by the DfES, and follow local regulations. Handle sources only with long tweezers.

- Line up the source and the GM tube with a gap of less than 1 cm between them.
- Measure the count recorded in 1 minute.
- Put the sheet of paper between the tube and source, and measure the count in 1 minute again.
- Replace the paper with the aluminium plate and count again for 1 minute.
- Finally measure the count with the piece of lead blocking the radiation. (You will have to move the GM tube to get it in, but the extra air gap will not absorb much gamma radiation.)

If the change in the count using different absorbers is small, take extra check readings (see the *Sample readings*).

Analysis

1 Record your measurements in a table.
2 Use your results to deduce the type of radiation coming from the source. Assume that a thin sheet of paper will absorb alpha particles but not stop beta particles. Assume also that 3 mm of aluminium will stop beta particles but not block gamma ray photons. (Note that the GM tube you are using may not be able to detect alpha particles at all. If this is the

case, 'no change' with a paper absorber does not mean that alpha particles are not being emitted by the source.)

Sample readings

The source was radium-226.
Average background count = 19 minute^{-1}.

Counts per minute

no absorber	79 909	78 115	79 536	79 769	average 79 332	std 715
0.2 mm paper	67 533	67 690	66 907	68 007	average 67 534	std 400
3 mm aluminium	3050					
3 cm lead	674					

Decide whether any alpha radiation was present when the sample readings were taken.
Explain why it was only necessary to test once for beta and gamma radiations.
Explain what the test with the lead absorber tells us about the gamma ray photons.

Evaluation

If the readings with and without an absorber are very close it can be difficult to decide whether there is a significant difference between them, especially since repeat readings will show a random variation. To check, subtract the standard deviation from the higher reading and add the standard deviation to the lower one. If there is still a difference it is likely that the original readings do represent a significant change.

Computer simulation

Identifying radiation

Aim **To identify the type of radiation from its penetrating power.**

Apparatus

- computer running the program 'Decay' from the CD

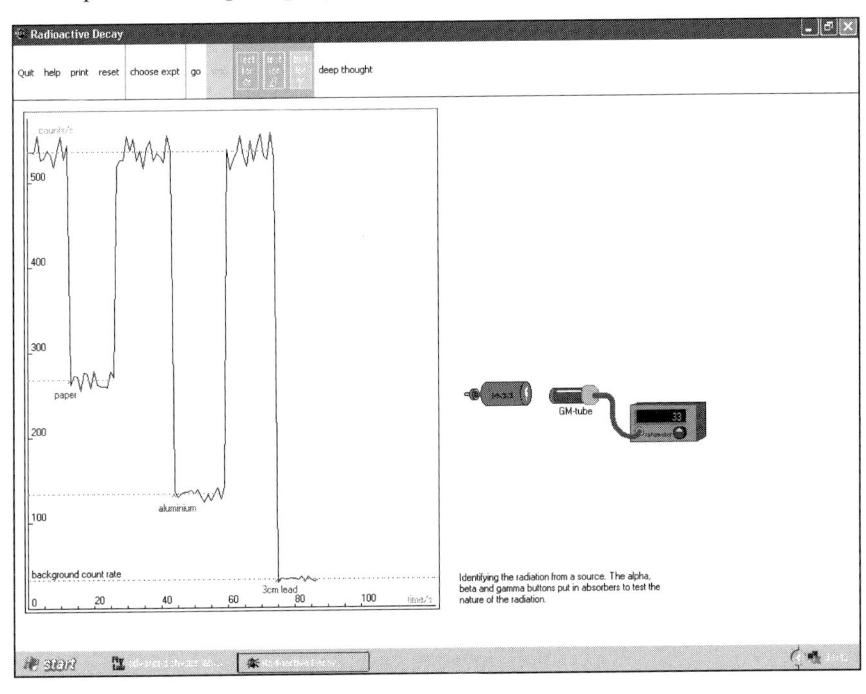

Plan

- Choose Experiment B.
- Click 'go', and the ratemeter reading (count/s) starts to be recorded.
- Put each absorber in turn into the beam of radiation by clicking 'test for α' (click again to remove the absorber), 'test for β', etc.
- Print a copy of the graph. Ratemeter readings will also be printed.

Analysis If paper can reduce the count rate, there must be alpha radiation present. There could also be beta and gamma. If aluminium causes a further drop in the count rate, beta radiation must be present. If the count rate is still above the background level, gamma rays must also be present. The lead confirms this if it reduces the count rate further.

Write a paragraph explaining, in your own way, the logic used in this test to identify the type of radiation that is present.

15.2 The nucleus

Computer simulation: Nucleus

(There are no lab experiments for this topic.)

Computer simulation: Nucleus

Aim **To investigate the scattering of alpha particles by the nucleus of an atom.**

Apparatus

- computer running the program 'Nucleus' from the CD

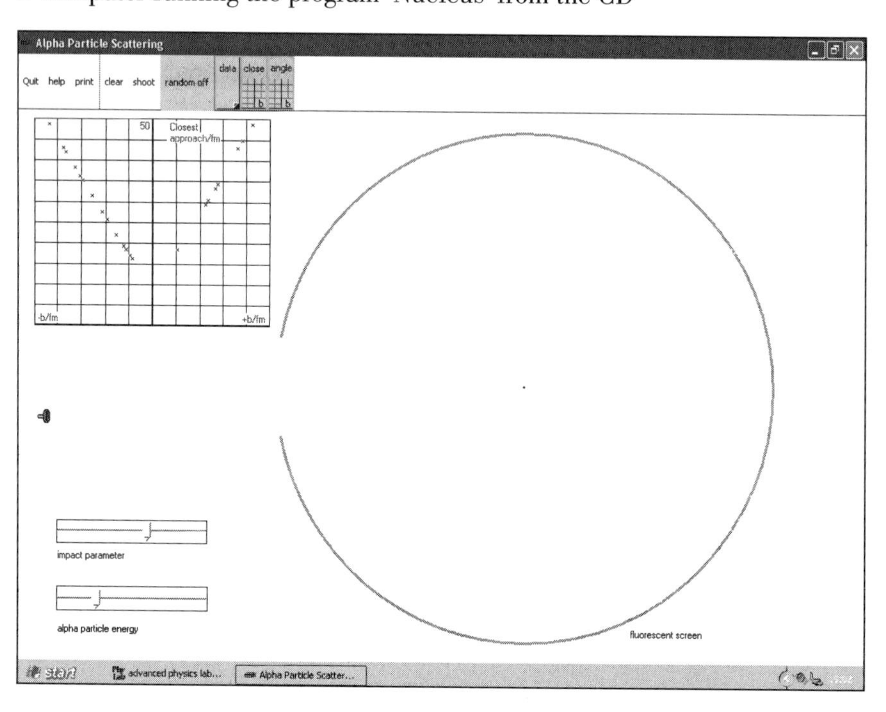

Plan

- Click 'shoot'. An alpha particle flies out from its source and is deflected by the repulsive force from the positive nucleus.
- Change the 'impact parameter' using the slider, so that the alpha particle is aimed at a different distance from the nucleus, and shoot again. The closest distance of approach and angle of deflection are calculated automatically (click on 'data').
- Continue until you have enough data to plot the graphs. (You can use 'random on' and 'shoot' to get continuous shots with random impact parameters.)
- Investigate the effect of alpha particle energy on how close the alpha particle gets to the nucleus.

Analysis

Write a general description of your observations, explaining how this experiment is able to give an estimate of the size of the atomic nucleus.

15.3 The half-life of a radioactive element

Preliminary work: Finding the half-life of 'radioactive dice'
Computer simulation: Half-life of radon
Experiment 1: Measuring the half-life of radon
Experiment 2: Measuring the half-life of protactinium

Preliminary work — Finding the half-life of 'radioactive dice'

Apparatus

- 100 or more dice (or small wooden cubes with one face coloured black)
- tray with sides

Plan

- Choose a number for the dice, or the coloured face of the cubes, to represent one that has 'decayed'.
- Count the original number of dice, shake them and throw them into the tray.
- Remove the 'decayed' dice, count and record the number left and shake them again.
- Shake on until all the dice have 'decayed'.

Analysis

1. Put your results into a table and plot a graph of 'number left' against 'number of throws'.
2. Draw a best-fit smooth curve through the points, remembering that the scatter is due to the random nature of your readings.
3. 'Half-life' in this simulation is the number of shakes it takes to reduce the original number of dice by half. Read the 'half-life' from the graph.

Alternative analysis The theoretical equation for an exponential decay is:

$N = N_0\, e^{-\lambda t}$ N_0 = number at time 0
N = number at time t
λ = decay constant

Taking logs to base e:

$\ln (N_0/N) = \lambda t$

So a graph of $\ln (N_0/N)$ against t will be a straight line if the decay is exponential, and its gradient will be λ.

1 Plot ln (original number/number left) against 'number of throws' and calculate the gradient, λ, of the best-fit line.
2 Calculate 'half-life' from:

half-life = $0.693/\lambda$

Sample readings

Dice showing '6' were counted as having decayed

no. of shakes	0	1	2	3	4	5	6	7	8	9	10	11	12	13	14	15
no. of dice left	400	345	282	229	202	185	159	144	123	102	90	80	61	51	45	38

no. of shakes	16	17	18	19	20	21	22	23	24	25	26	27	28	29	30	31
no. of dice left	28	25	19	14	13	11	9	8	7	6	6	6	5	4	2	0

Dice showing '6' and '3' were counted as having decayed

no. of shakes	0	1	2	3	4	5	6	7	8	9	10
no. of dice left	400	274	197	141	105	75	58	43	28	19	12

Plot number left (N) against number of shakes (t) for both sets of sample results on the same axes. Read off the half-life for each.

Answers 4.0 shakes; 2.0 shakes.

Plot $\ln (N_0/N)$ against t, as suggested in the *Alternative analysis*, for both sets of sample results. Draw the best-fit straight lines. Measure the decay constant (= gradient) for each.

Answers 0.17; 0.34.

For each, calculate the 'half-life' from the decay constant (half-life = 0.693/decay constant).

Answers 4.1 shakes; 2.0 shakes.

Computer simulation

Half-life of radon

Aim **To measure the half-life of radon (a radioactive gas) from its decay curve.**

Apparatus

- computer running the program 'Decay' from the CD

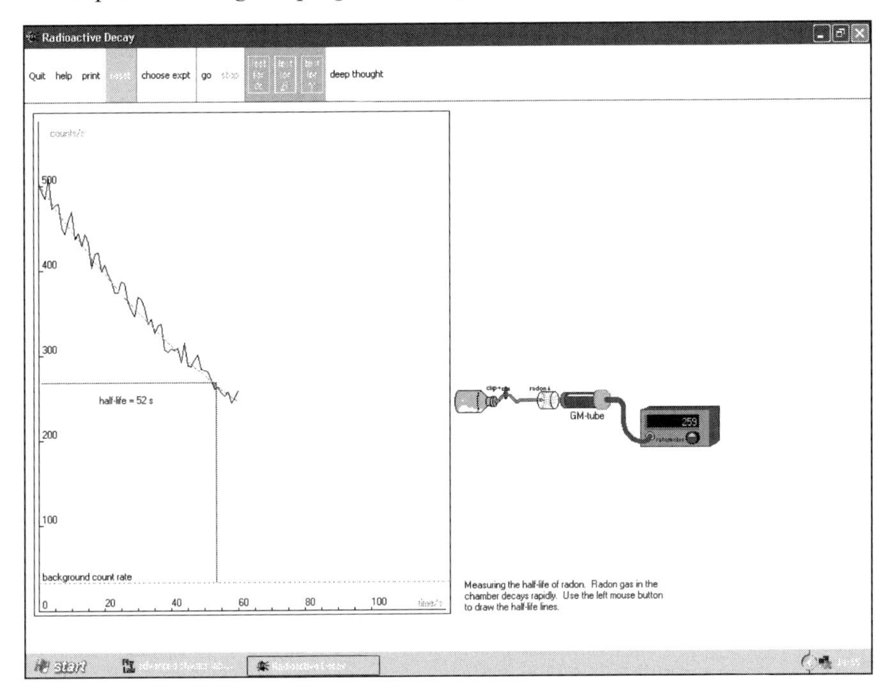

Plan

- Choose Experiment C.
- When you click 'go', radon gas is squeezed into a small chamber and a clip is closed over the tube. This isolates the gas from its 'parent' and it begins to decay.
- Leave the experiment to run, and watch how the count rate drops at a decreasing rate. Note that the readings vary in a random manner about a smoothly decreasing mean.
- After the activity has dropped to below half, click on the curve and construction lines are drawn to read off the half-life. You can do this a number of times.
- Print a copy of the graph and readings.

Experiment 1 Measuring the half-life of radon

Apparatus

- radon generator (see below) attached by a tube to a small chamber with a thin plastic window
- clip to close the tube
- GM tube connected to a scaler with power supply for the tube
- stop clock or watch

The radon generator is a soft polythene bottle that contains a sealed quantity of thorium oxide. Over time, the thorium decays and produces radon-220 gas which builds up in the bottle.

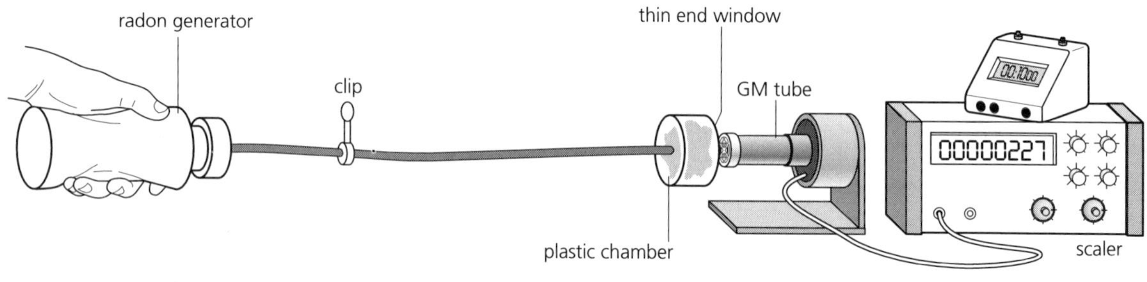

Figure 15.3

Plan **Take great care when handling radioactive materials**, especially when they are gaseous (as here) or in powder form. Follow the Local Authority regulations for the use of a radon generator.

- Support the chamber with its window as close as possible to the GM tube (see Figure 15.3).
- Measure the background count (see p. 183) with the bottle and chamber in position.
- Open the clip on the tube, squeeze the bottle firmly to force radon into the chamber, close the clip and immediately start the counter and stop clock.
- After 10 s record the count and zero the scaler but keep the clock running.
- At 30 s record the count in 10 s again, and then every 30 s for 3 minutes.

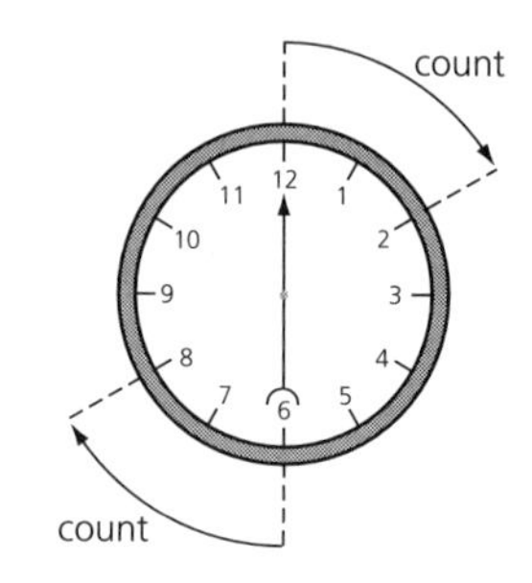

Figure 15.4

Analysis

1. Put your results into a table (remembering to deduct the background count) and plot a graph of the count/10 s against time.
2. Draw a best-fit smooth curve through the points, remembering that the scatter is due to the random nature of the decay.
3. Construct lines on the graph to read off how long it takes the count to halve, to go from a half to a quarter and from a quarter to an eighth. Work out the average as your best estimate for the half-life of radon.

Sample readings

time /s	0	30	60	90	120	150	180
count /10 s	200	138	100	71	50	35	25

The background count has been deducted from these readings.

Plot the graph suggest in the *Analysis* for the sample readings and read off an average value for the half-life of radon.

Answer ~57 s.

Plot the log graph suggested in *Improving the analysis* below. Measure the decay constant (λ) from the gradient and calculate the half-life ($= 0.693/\lambda$).

Answers ~0.012 s^{-1}; ~58 s.

Evaluation It is impossible to do check readings, and running readings have to be taken of count and time. The count is not a constant quantity because it depends on the random nature of the decay of the radon nuclei. These introduce uncertainty in the choice of a best-fit line that adds to the normal uncertainties of reading values from a graph.

The easiest way to estimate the uncertainty in half-life is from the difference in the half-, quarter- and eighth-life values. This could be 12 s in 60 s or about 20%.

Improving the analysis To improve the result plot a graph of ln (N_0/N) against t, where N_0 is the count/10 s at time zero and N the count/10 s at time t. Use the gradient to measure the decay constant and half-life (see the *Alternative analysis* on p. 188).

Experiment 2 Measuring the half-life of protactinium

Apparatus

- GM tube
- scaler with power supply for the GM tube
- stop clock or watch
- protactinium generator (see below)
- clamp and stand

Protactinium (Pa) has a half-life of about 70 s. It can be produced by the decay of ^{238}U:

$$^{238}U \rightarrow {}^{234}Th \rightarrow {}^{234}Pa \rightarrow {}^{234}U$$

Pa is soluble in an organic solvent but the other products are not. To make a protactinium generator, dissolve 1 g of uranyl nitrate in 3 cm^3 of water and 7 cm^3 concentrated HCl. Put into a small thin-walled plastic bottle with a secure lid. Add 10 cm^3 of amyl acetate and close the lid.

Plan **Take great care with this experiment.** Keep the bottle of radioactive liquids as far from you as you can. Follow the Local Authority regulations for the use of open radioactive sources.

- Clamp the protactinium generator with its upper (organic) layer of liquid at the height of the GM tube (see Figure 15.5).

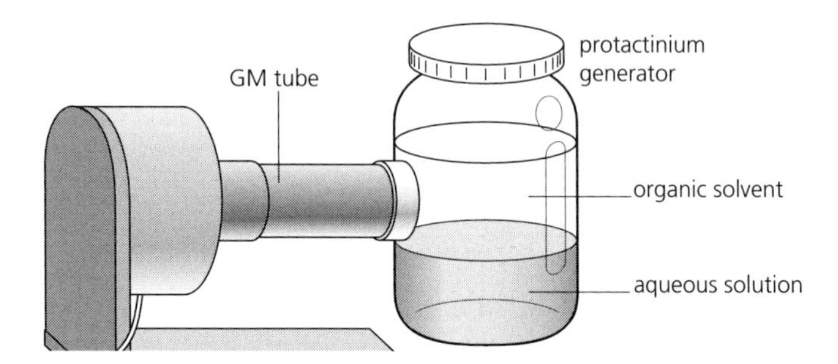

Figure 15.5

- Prepare the GM tube and scaler. Make sure they are working and measure the radiation count with the bottle in position. Do this by recording the count every 20 s for 2 minutes. Find the average count in 20 s.
- Release the bottle and shake vigorously for 30 s. This dissolves the protactinium from the aqueous solution into the organic solvent.
- Put the bottle back in the clamp and push it up close to the GM tube. As soon as the oily organic liquid starts to separate out as an upper layer, start the scaler and the stop clock.
- After 20 s record the count reading but keep the clock and the scaler running.
- At 40 s record the count reading again, and then every 20 s for 7 minutes.

Analysis

1. Put your readings into a spreadsheet or table. You will need to work out the count in 20 s and the 'corrected count' in 20 s (count/20 s − average count/20 s before shaking). See the *Sample readings* below.
2. Plot a graph of the corrected count/20 s against time.
3. Draw the construction lines you need to read off the time it takes the count to halve.
4. For greater precision find the decay constant and half-life by plotting $\ln (N_0/N)$ against t (see *Alternative analysis* on p. 188). The points may become very scattered when the count rate is low and should be ignored when choosing the best-fit line.

Sample readings

Before shaking

time /s	0	20	40	60	80	100	120	140	
total count	58	101	157	203	257	291	351	396	
count /20 s		43	56	46	54	34	60	45	average 48

After shaking

time /s	0	20	40	60	80	100	120	140	160	180	200
total count	171	334	471	610	725	818	912	990	1049	1122	1175
count /20 s	171	163	137	139	115	93	94	78	59	73	53
corrected /20 s	123	115	89	91	67	45	46	30	11	25	5

Plot both graphs suggested in the *Analysis* above for the sample readings. Use the graphs to obtain values of the half-life and decay constant of protactinium.

Answers

~80 s; 74 s; 0.0094 s^{-1}.

Skill level (Analysing)

A: I recorded the actual readings of the scaler in a headed table. I calculated the count over successive 20 s intervals. I calculated the corrected count. I found half-life from a graph of count against time. I found half-life from the gradient of a log graph.

All but one of the above = B; all but two = C; all but three = D; all but four = E.

15.4 Beta and gamma radiation

Full investigation 1: How far can beta radiation penetrate aluminium?
Full investigation 2: Does gamma radiation from a point source follow the inverse square law?

Local Authority regulations on the use of sealed radioactive sources should be adhered to in the practical work in this topic.

Full investigation 1 How far can beta radiation penetrate aluminium?

Apparatus

- GM tube
- scaler with power supply for the GM tube
- sealed source of beta radiation (e.g. strontium-90) in a holder
- long tweezers for handling the source
- aluminium plates (e.g. from a radioactivity kit) with a range of thicknesses, marked with their 'surface density', i.e. mass/area (squares of cardboard can be used instead because the ability of materials to stop beta radiation depends largely on the surface density and not the material)
- stop clock or watch

Plan

Take great care to keep the source at arm's length, always handle with long tweezers, and never point it at anyone's eyes or sex organs.

- Measure the average background count per minute (see p. 183).
- Line up the source and GM tube, leaving space between them for the aluminium plates.
- Put the thinnest aluminium plate in the gap and measure the count for 1 minute.
- Record or measure the surface density (mass/area) of the plate.
- Repeat for all of the plates. You can also combine the plates together to make thicker plates.

Skill level (Implementing)

A: I adjusted the GM tube and scaler so that they recorded the background radiation. I measured the background count/minute a number of times and found the average. I used long tweezers to put the beta source in its holder and kept it well away from myself and others. I kept the source and GM tube a fixed distance apart. I measured the count/minute for a range of metal sheets until it reached background level.

All but one of the above = B; all but two = C; all but three = D; all but four = E.

Analysis

1. Make a table for your results with two rows: surface density of the plate, and count/minute.
2. Plot a graph to show how the count rate decreases with surface density of the plate.
3. Dot in a horizontal line showing the background count rate. Extrapolate your graph to meet this line.
4. Read off from the graph the surface density (D) that is needed to stop all of the beta particles from the source (i.e. that would reduce the count rate to background level).

The maximum energy of the beta particles is given by this empirical formula:

$E = (D + 0.133)/0.542$ (D is in g cm^{-2}; E is in MeV)

5. Calculate the maximum energy of the beta particles coming from the source.

Sample readings Average background count = 18/minute
Source: strontium-90

count /minute	73505	51177	43287	29915	5471	1344	275	76	72	52
surface density /mg cm^{-2}	0	94	141	248	543	698	894	1592	2135	2618

Plot the graph suggested in the *Analysis* for the sample readings and calculate a value for the maximum energy of the beta particles.

Answer ~1.9 MeV.

Full investigation 2 Does gamma radiation from a point source follow the inverse square law?

Apparatus

- sealed gamma source (e.g. cobalt-60) in a holder
- thin sheet of aluminium
- protective lead shield if available
- long tweezers
- GM tube in a holder
- scaler with power supply for the GM tube
- stop clock or watch
- metre rule (or base board from a radioactivity kit)

Plan **Take great care with the gamma source.** The radiation can travel some distance and penetrate deep into your body. Keep it at arm's length and stand at the side of or behind the source. Use a protective lead shield if you have one. The risk of harm from a school gamma source is low but it is good to practise the precautions necessary with stronger sources.

- Move the source at least 2 m away from your detector and measure the average background count (see p. 183).
- Place the aluminium sheet in front of the source to absorb any beta radiation.
- Place the window of the GM tube 1 cm from the source and measure the count in 30 seconds. You may find you get a larger reading if the GM tube is placed at right angles to the beam. (The gamma rays interact with the side walls of the tube, knocking electrons into the tube.)
- Move the source away from the tube in 1 cm steps and record the count/30 s for each distance.

Analysis The *inverse square law* states that the intensity of radiation passing normally through an area is inversely proportional to the square of its distance from its source. For this investigation, C, the count in 30 s, represents the intensity of the radiation and $(d + e)$ represents the distance from the source. Here d is the measured distance and e is a correction needed, because the exact positions of the source and of the detection in the GM tube are not known.

Using the inverse square law:

$$C \propto 1/(d + e)^2$$

$$\sqrt{C} \propto 1/(d + e)$$

$$1/\sqrt{C} = k(d + e) \quad \text{where } k \text{ is a constant}$$

If this equation applies, a graph of $1/\sqrt{C}$ against d should be a straight line with a negative intercept on the x-axis equal to the unknown distance e.

1 Put your results into a spreadsheet. Use it to calculate $1/\sqrt{C}$ and plot the graph. Choose the best-fit line and read off the value of e. Add this correction to your values of d and replot the graph. If the points now lie on a straight line through the origin, your results verify that gamma radiation does follow the inverse square law.
2 A second check is to use a column in the spreadsheet to calculate $C \times (d + e)^2$. The values will tell you whether intensity and $1/(\text{distance})^2$ are proportional within the precision of your readings. (See the *Sample readings* overleaf.)
3 Write a conclusion, saying whether you think your results do or do not show that gamma radiation from this source follows the inverse square law.

Sample readings Source: cobalt-60
Average background count/30 s = 8

count /30 s	C (corrected count /30 s)	1/√C	d /cm	(d + e) /cm	C × (d + e)²
633	625	0.040	1.0	2.0	2500
306	298	0.058	2.0	3.0	2682
166	158	0.080	3.0	4.0	2528
112	104	0.098	4.0	5.0	2600
89	81	0.111	5.0	6.0	2916
76	68	0.121	6.0	7.0	3332
53	45	0.149	7.0	8.0	2880
33	25	0.200	8.0	9.0	2025
29	21	0.218	9.0	10.0	2100
27	19	0.229	10.0	11.0	2299
					average 2586
					std 375
					% uncertainty 14

Correction e /cm = 1.0

Plot a graph of $1/\sqrt{C}$ against d for the sample readings. Draw the best-fit line, ignoring any anomalous points. Confirm that the correction $e = 1.0$ cm. Add this correction to values of d and plot the graph of $1/\sqrt{C}$ against $(d + e)$. Decide whether the points justify drawing a straight line through the origin.

Answer Yes, just about, if the three points with the lowest count rates are ignored.

Skill level (Analysing)

A: I entered the count and distance readings into a spreadsheet. I used a formula to calculate the corrected count and 1/√C. I used the program to plot a graph of 1/√C and distance. I entered formulae to check for proportionality. I wrote a conclusion about the inverse square law.

All but one of the above = B; all but two = C; all but three = D; all but four = E.

Evaluation A large cause of uncertainty is the variation of the count rate due to the random nature of the decay of the cobalt. Taking a number of readings for a fixed distance will give you an idea of the range of the uncertainty (from ~3% to ~10%, depending on the count rate). This means you cannot give a precise answer and your conclusion can only be about 90% certain.

Uncertainty arises in the value of d, but this is allowed for in the *Analysis*.

Improving the plan

- Increase the time for which you measure the count as the count rate gets lower. This will reduce the % uncertainty in the low readings.
- Position the GM tube sideways on to the radiation to see if this increases the count rate, or try wrapping the sides of the GM tube in copper foil.

Index

Entries in *italic* refer to the programs on the accompanying CD; all page numbers refer to this book.